DRAWING

for Civil Engineering

Second Edition

Jan A van der Westhuizen

Drawing for Civil Engineering
First published 2000
Second Edition 2012

Juta and Company Ltd
First floor, Sunclare Building
21 Dreyer Street, Claremont, 7708

©Juta and Co Ltd, 2012

ISBN 978 0 70218 873 2

This book is copyrighted under the Berne Convention. In terms of the Copyright Act 98 of 1978, no part of this book may be reproduced or transmitted in any form or by any means, including photocopying, recording, or by any information storage and retrieval system, without permission in writing from the publisher.

Project managed by Corina Pelser
Editing by Linette Downs-Webb
Proofreading by Simone van der Merwe
Indexing by Jennifer Stern
Typeset by Trace Digital Services
Cover design by Marius Roux
Printed and bound in South Africa by Formeset Print, Cape Town

The authors and the publisher have made every effort to obtain permission for and acknowledge the use of copyright material. Should any infringement of copyright have occurred, please contact the publisher, and every effort will be made to rectify omissions or errors in the event of a reprint or new edition.

Contents

MODULE 1 REINFORCED CONCRETE ... 1

Requirements for detailing reinforced concrete drawings ... 1
Module outcomes ... 1
Terms ... 2

Unit 1 Introduction and standard tables ... 4

1.1 Reinforced concrete ... 4
    1.1.1 Simple theory ... 4
1.2 Detailing of reinforcement ... 5
    1.2.1 Beams ... 5
    1.2.2 Columns ... 5
    1.2.3 Slabs ... 5
    1.2.4 Stirrups or links ... 6
1.3 General principles for drawing ... 8
1.4 Bar schedules (or bending schedules) ... 12
1.5 Types of drawings used for reinforced concrete ... 15
1.6 Summary ... 15
Self-evaluation ... 16

Unit 2 Foundations and columns ... 17

2.1 Introduction to foundations ... 17
2.2 Types of bases ... 17
    2.2.1 Spread footings ... 17
    2.2.2 Pile footings ... 17
2.3 Introduction to columns ... 19
2.4 Method of detailing columns ... 19
2.5 A closer look at footings and columns ... 20
    2.5.1 Isolated footings ... 20
    2.5.2 Combined footings ... 21
2.6 Summary ... 24
Solved examples ... 24
Activity 2.1 ... 33
Activity 2.2 ... 34
Self-evaluation ... 35

Unit 3 Beams and slabs ... 36

3.1 Introduction to beams ... 36
3.2 Detailing beams ... 38
3.3 Introduction to slabs ... 44
3.4 Detailing floor slabs ... 47
3.5 A practical approach ... 50
3.6 Summary ... 54
Solved examples ... 54
Activity 3.1 ... 77
Activity 3.2 ... 78
Activity 3.3 ... 78
Activity 3.4 ... 79
Activity 3.5 ... 80

Drawing for Civil Engineering

		Self-evaluation	82
Unit 4		**Using Computer Aided Concrete Training**	**83**
	4.1	Introduction to COMPACT	83
		4.1.1 Advanced Design of Reinforced Concrete Structures	84
		4.1.2 Design of Reinforced Concrete Structures	85
		4.1.3 Buildability	86
		4.1.4 Conceptual Design of Concrete Structures	87
		4.1.5 Concrete as a Material (including Mix Design)	88
		4.1.6 Concrete Bridges	89
		4.1.7 Concrete Site Practice	90
		4.1.8 Drawing and Detailing of Concrete Structures	91
		4.1.9 Foundations and Retaining Walls	92
		4.1.10 Precast Concrete Structures	93
		4.1.11 Pre-stressed Concrete Structures	94
	4.2	Downloading COMPACT	95

MODULE 2 STRUCTURAL STEELWORK 97

Requirements for detailing structural steel drawings 97
Module outcomes 97
Terms 98

Unit 5		**Tables**	**100**
	5.1	Introduction	100
	5.2	Standard steel tables	100
	5.3	Bolted connections	101
	5.4	Backmark	118
	5.5	Dimensioning of holes	120
	5.6	Symbols	121
	5.7	Holding down bolts (HD bolts)	122
	5.8	Welded connections	123
		5.8.1 Types of welds	124
		5.8.2 Symbols for welds	124
		5.8.3 Sizes of welds	125
	5.9	Summary	126
		Activity 5.1	126
		Activity 5.2	126
		Self-evaluation	128
Unit 6		**Base-to-column connections**	**129**
	6.1	Bases	129
	6.2	Columns	130
		6.2.1 Types of columns	130
		6.2.2 The grid system	131
		6.2.3 Column splices	131
	6.3	Summary	135
		Activity 6.1	135
		Self-evaluation	137
Unit 7		**Beam-to-column connections**	**138**

		7.1	Beams	138
			7.1.1 Notching of beams	138
			7.1.2 Eccentric connections	142
		7.2	Beam-to-column connections	142
		7.3	Summary	142
			Activity 7.1	142
			Activity 7.2	143
			Activity 7.3	144
			Self-evaluation	145
Unit 8	**Beam-to-beam connections**			**146**
		8.1	Beams	146
		8.2	Sequential system	146
		8.3	Splicing beams	146
		8.4	Ways of connecting beams to beams	149
		8.5	Summary	151
			Activity 8.1	151
			Self-evaluation	152
Unit 9	**Roof structures**			**153**
		9.1	Introduction to roofs	153
		9.2	Roof trusses	153
		9.3	Lattice girders	160
		9.4	Portal frames	165
		9.5	Roof systems	169
			9.5.1 A typical truss and purlin system	169
			9.5.2 Lattice girders	170
			9.5.3 Portal frames	171
		9.6	Summary	172
			Activity 9.1	172
			Activity 9.2	173
			Activity 9.3	174
			Activity 9.4	175
			Activity 9.5	176
			Self-evaluation	177

MODULE 3 SURVEYING 179

Introduction to surveying drawing for civil engineering 179
Module outcomes 179
Terms 180

Unit 10	**Introduction to surveying**		**181**
	10.1	Introduction	181
	10.2	Grid lines (revision)	181
	10.3	Plotting spot-heights (revision)	183
	10.4	Contouring (revision)	183
		10.4.1 Draw in contours	183
		10.4.2 Contour values	185
		10.4.3 Contour characteristics	186
		10.4.4 Physical features	186
		10.4.5 Conventional symbols	186

Drawing for Civil Engineering

    10.5 Step-by-step method of drawing grids and contours 189
        10.5.1 Necessities on any plan.. 189
        10.5.2 Orientation of the plot and plotting grid lines........................ 189
        10.5.3 Plotting sequence 197
    10.6 Summary 199
    Solved example 199
    Activity 10.1 201
    Self-evaluation 202

Unit 11 Positioning of structures on a site plan 203

    11.1 Introduction... 203
    11.2 Cut and fill... 203
    11.3 The mass haul diagram (MHD).. 203
        11.3.1 General 203
        11.3.2 Preparation... 204
        11.3.3 Bulking and shrinkage 205
        11.3.4 Properties of the MHD 205
        11.3.5 Balancing procedure... 206
    11.4 Summary... 207
    Solved examples 207
    Activity 11.1 217
    Activity 11.2 218
    Self-evaluation 218

Unit 12 Introduction to CAD in surveying... 220

    12.1 Introduction to SURPAC SURVEY SOFTWARE............................. 220
        12.1.1 Co-ordinate File 220
        12.1.2 Opening an existing or creating a new Co-ordinate File 220
        12.1.3 Displaying the Co-ordinate File 221
        12.1.4 Auto-Compilation of an Engineering Format Beacon
               Description List 221
        Activity 12.1 222
        12.1.5 Printing the Co-ordinate File 223
        12.1.6 Selecting the 'Set Text File (*.vpg) as Printer' option from
               the File Menu 223
    Activity 12.2 224
    12.2 The General CAD Construction/Edit/Plotting Program 225
    12.3 CAD Plotting Sheet Construction/Editing 226
        12.3.1 Line and Arc Drawing 226
        12.3.2 General Text Writing 228
    Activity 12.3 229
    12.4 The SURPAC ENGINEERING Module Applications 229
        12.4.1 Horizontal Curve and Straight Alignment............................... 229
        12.4.2 Cross-section Creation/Editing/Plotting................................. 232
        12.4.3 Longitudinal Section Creation/Editing/Plotting.................... 235
        12.4.4 Sectional Volumes 237

Appendix... 240

Acknowledgements 248

Index.. 255

Note to the student

This book provides practical and up-to-date information on drawing for civil engineering. Even though explanations seem to be mechanical, the aim is for students to use it in conjunction with CAD and it can be applied with any available CAD software, such as AutoCAD, AlleyCAD and the like. Aimed at second-year students, it covers the fundamentals of drawing, as well as draughting practice and conventions. Although *Drawing for Civil Engineering* is designed for use by students, it will also be a useful reference source for educators, draughtspeople, practising engineers, fabricators and contractors in the field of civil engineering.

Designing and detailing complicated structures for specific contracts and research assignments is a complex task. This publication is unique in that it covers a wide range of topics — reinforced concrete, structural steelwork and surveying — in one book. It helps to explain the importance of the role of drawings within the civil engineering construction process as a whole, and explains why it is necessary to adopt standard methods of presentation for drawings. The language used in the text is simple, conversational English, with technical terminology and difficult concepts explained throughout the book.

Drawing for Civil Engineering consists of three modules. Each module starts with a list of study objectives or outcomes. These outcomes set out what you should be able to do at the end of each module.

The text is set out in such a way that you should be able to work through the book by yourself. New concepts are explained and reinforced by providing examples with solutions to work through. In addition, many figures are used throughout the text to aid understanding and clarify concepts. Because this is a problem-solving course, there are also many activities for you to work through. These activities allow you to make sure that you understand the work you have covered in a particular unit.

The summary at the end of each unit enables you to see at a glance what you should have learnt in that unit. The summary is followed by a section with self-evaluation questions to enable you to assess your understanding of the concepts discussed in that unit. Answers to self-evaluation questions appear at the end of each unit.

Icons

The icons used in *Drawing for Civil Engineering* are explained below.

 This is an activity icon. When you see this icon you will know that it is time to do something! The activities are enjoyable, and they help you to understand the subject and monitor your progress. Feel free to do the activities with a fellow student or group of students. The solutions to some activities are given in the text, but for most of the activities you will need to ask your lecturer to check your work.

 This is a terminology icon. Read the definitions of the terms carefully because the details are important.

 The take note icon appears alongside all the extremely important information.

 This is a self-evaluation icon. The self-evaluation questions at the end of each unit enable you to assess your understanding of concepts discussed in that unit.

 Reminder

Introduction to civil engineering drawing

In the civil engineering discipline, the designer of a structure must be able to communicate his or her design requirements to the contractor who will be building the structure. The most effective way to do this is for the designer to produce a set of drawings which clearly and unambiguously set out the structure and all the requirements for its successful completion. The drawings must be clearly and neatly set out so that there can be no doubt as to what is required. Dimensioning of the structure is of the utmost importance, and this must be done so that the contractor is in no doubt as to the location and shape of the structure. Errors in the drawings can lead to very costly remedial work on site, which might involve breaking down and rebuilding parts of the structure.

This book aims to enable effective communication through the medium of draughting. This is achieved by:

❐ introducing you to the art of producing civil engineering drawings in accordance with current practice and regulations, so that the drawings are legible and can be used by contractors and builders without having to refer back to you;
❐ introducing you to the art of interpreting and understanding drawings of a civil, surveying and architectural nature and to the terminology used in these fields; and
❐ guiding you into developing your individual draughting skills/style to obtain an optimum balance of presentation and speed.

Please note that the figures in this book are generally not drawn to scale. In the few cases where a figure is drawn to scale, the scale used is provided with the drawing.

This book is dedicated to my wife, Louisa,
and my two sons, Roan and Iwan.

J.A. van der Westhuizen

module 1

Reinforced concrete

The photograph shows the construction of the turbine house and auxiliary bay for six generating units at Kendal Power Station.

Photographer: J Jung
Reproduced by kind permission of Murray & Roberts Gillis Mason

Requirements for detailing reinforced concrete drawings

The detailing of the reinforcing is done according to SANS codes 282 and 0144.

Module outcomes

After studying this module, you should be able to use the different tables and shape codes to:
a. produce complete and neatly dimensioned concrete drawings of simple structures, such as:
   - reinforced concrete bases;
   - reinforced concrete columns;
   - reinforced concrete slabs;
   - reinforced concrete beams; and
   - reinforced concrete framed structures involving bases, columns, slabs and beams.
b. produce the relevant reinforcing drawings for these structures and provide the bending schedules.

Module 1 consists of Units 1, 2, 3 and 4.

Drawing for Civil Engineering

 The words below are printed in bold the first time they are used in the text.

Aggregate: Pieces of crushed stone or gravel used in making concrete.
Cement: A powdery substance made by calcining (reducing to calcium oxide by strong heat) lime and clay.
Concrete: A composition of aggregate, cement and water used for building.
Concrete cover: A predetermined layer of concrete to protect the steel bars from moisture that will rust the steel, and from the heat of any fire that could degrade the steel and lead to structural collapse (see Fig 1.4 on page 5).
High yield steel (as shown in Fig 1.1): Deformed (rough texture to ensure a better grip) steel bars of strength 450 MPa.

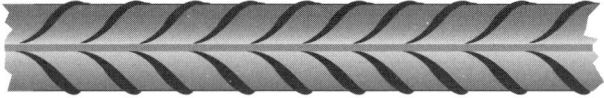

Fig 1.1 High yield steel bar

Lap lengths: Whenever we combine two pieces of steel, we do it in such a way that it overlaps to bond with the concrete without weakening the reinforcing at that point. The length to overlap we read directly from Table 1.1 on page 3. Depending on the nature of the stress, these laps can either be in compression or tension.

Fig 1.2 Mild steel bar

Reinforced concrete: Concrete with steel bars embedded to increase its tensile strength.
Reinforcement: The steel used to reinforce concrete.
Stirrups: Steel shaped like a stirrup to hold the longitudinal steel in place for columns and beams (see Fig 1.4 on page 5).
Forty-five degree (45°) dispersion angle: In footings, the load in the column is transferred into the base. This has the effect of punching a hole through the base. Failure due to this shear occurs approximately along 45° lines when footings are tested to destruction (see Fig 1.3).

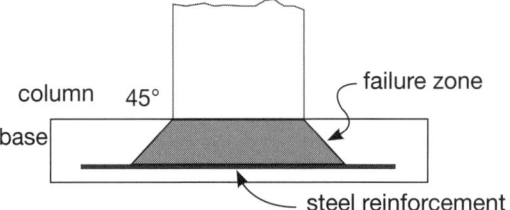

Fig 1.3 45° dispersion angle

Table 1.1 Minimum bond and lap lengths for fully stressed bars

1	2	3	4	5	6	7	8	9	10	11	12	13	14	15	16	17	18	19
Steel type	Stress classification	Concrete class	Length required (in diameters)*	Minimum bond length mm							Length required (in diameters)*	Minimum lap length mm						
				Diameter of bars mm								Diameter of bars mm						
				8	10	12	16	20	25	32		8	10	12	16	20	25	32
High yield deformed steel (450 MPa)	Tension	20	44	350	440	530	700	880	1100	1410	55	440	550	660	880	1100	1380	1760
		30	34	270	340	410	540	680	850	1090	43	340	430	520	690	860	1080	1380
		40	29	230	290	350	460	580	730	930	36	350	400	450	580	720	900	1150
	Compression	20	30	270	300	360	480	600	750	960	30	310	350	390	480	600	750	960
		30	23	180	230	280	370	460	580	740	23	310	350	390	470	550	650	790
		40	20	160	200	240	320	400	500	640	20	310	350	390	470	550	650	790
	Stirrups in shear (bent-up bars as for tension)	20	42	340	420	500	670	840	1050	1340	52	420	520	620	830	1040	1300	1660
		30	32	260	320	380	510	640	800	1020	40	350	400	480	640	800	1000	1280
		40	27	220	270	320	460	540	680	860	34	350	400	450	550	680	850	1090
Plain round mild steel (250 MPa)	Tension	20	45	360	450	540	720	900	1130	1440	45	360	450	540	720	900	1130	1440
		30	36	290	360	430	580	720	900	1150	36	350	400	450	580	720	900	1150
		40	29	230	290	350	460	580	730	930	29	350	400	450	550	650	780	950
	Compression	20	30	240	300	360	480	600	750	960	30	310	350	390	480	600	750	960
		30	24	190	240	290	380	480	600	770	24	310	350	390	470	550	650	790
		40	20	160	200	240	320	400	500	640	20	310	350	390	470	550	650	790
	Stirrups in shear (bent-up bars as for tension)	20	45	360	450	540	720	900	1130	1440	45	360	450	540	720	900	1130	1440
		30	36	290	360	430	580	720	900	1150	36	350	400	450	580	720	900	1150
		40	29	230	290	350	460	580	730	930	29	350	400	450	550	650	780	950

* or 20 d + 150 mm in compression;
or 25 d + 150 mm in tension and shear, whichever is greatest

Notes
1 Figures for 'length required (in diameters)' have been rounded off, as have equivalent bond and lap lengths.
2 No reduction in bond lengths has been allowed for in the table for the anchorage values of bends and hooks.
(These values are given in 6.4.4. of SABS 0144:1995.)

Unit 1 Introduction and standard tables

1.1 Reinforced concrete

1.1.1 Simple theory

*Do you know what **reinforced concrete** is?* Let us start by explaining what concrete is. Concrete is a hard, durable material made of graded **aggregate**, bound together with a paste **cement** and water. It is very strong in compression, but relatively weak in tension. This means that when concrete is fully supported it can take heavy loads, but the moment it is allowed to bend under its own weight, it will break.

By contrast, steel is very strong in tension. The solution is to combine the two to give a structure that is strong in both tension and compression. This is what is meant by reinforced concrete — a combination of concrete and steel. Think about the relationship between your skeleton and your flesh. The bones inside your body are actually responsible for the shape of your body, and your flesh covers the skeleton to provide a smooth protected appearance. In the same way, the steel functions as the 'skeleton' of a structure, while the concrete is the 'flesh' of a structure.

Reinforcement is needed to counteract tensile failure. As a point of interest, reinforcement may also be needed for reasons other than strength. In the case of a beam, for example, the longitudinal steel bars forming the main reinforcement are placed in the bottom, and transverse steel bars, called **stirrups**, are placed vertically to counteract shearing forces. However, these stirrups (links) would collapse when the concrete is placed unless they are made into a rigid 'cage'. Adding hanger bars, which complete the cage and hold the stirrups during the concreting operation, achieves this. For more clarification on this concept, study Fig 1.4 on page 5 and Fig 3.5 on page 40.

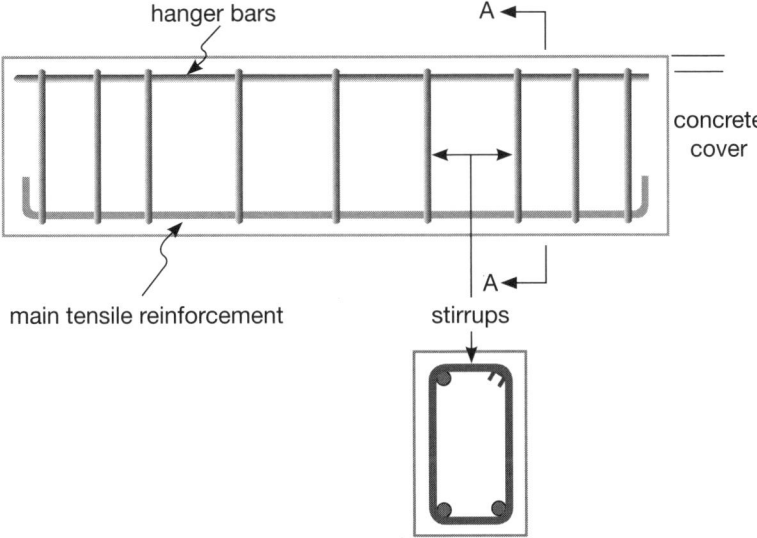

Fig 1.4 Basic beam reinforcement

1.2 Detailing of reinforcement

Detailing involves schematically describing the way in which you are going to place the steel. The purpose of detailing reinforced concrete is to convey to the contractor the information needed to get each and every steel bar fixed in its correct position. To draw every bar would produce a very complicated drawing that would be difficult to read. (This statement will become clear when we deal with reinforcement in the units to follow.) To standardise detailing, the following rules apply.

1.2.1 Beams

All main bars should be drawn in full in elevation and shown in cross-section.

1.2.2 Columns

Only one bar of each type needs to be shown in full elevation, but all bars should be shown in cross-section.

1.2.3 Slabs

Only one bar of each type should be shown in full in plan, but all should be shown in cross-section. The exception for slabs is where bars are staggered or alternate bars are reversed. In these cases, a pair of bars should be drawn in full.

1.2.4 Stirrups or links

Only one stirrup or set of stirrups in a beam and only one link or set of links in a column needs to be shown in full. This will become clear in the units to follow. Reinforcing bars are shown as single lines, 0,7 mm thick. Bars are described in a standard way using recognised abbreviations in the following sequence:
- number of bars required;
- type of steel;
- size of bar;
- identification mark number;
- spacing — centre to centre;
- location; and
- comment, if any.

We abbreviate the following as such:

For type of steel:
- R — mild steel; or
- Y — **high yield steel**.

For location:
- T — top;
- B — bottom;
- EW — each way;
- ABR — alternate bars reversed (every second bar's bent end shows in the opposite direction to the other bar);
- STGD — staggered (two bars offset by distance *d*);
- ALT — alternate different bars; and
- TOG — together.

Examples of the use of some symbols are given in Fig 1.5 on page 7.

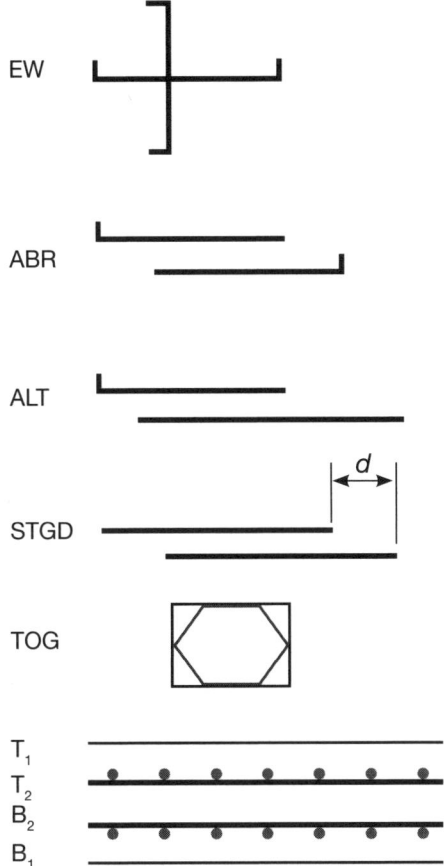

Fig 1.5 Example of symbols

If a bar description reads 24 Y20-03-200B ABR, for example, it means 24 high yield bars, 20 mm in diameter, mark 03 at 200 mm centre to centre in the bottom of the slab with alternate bars reversed (see Fig 1.5). When spacing reinforcement, you allow the builder space to insert his poker (vibrator) to compact the concrete without touching the reinforcement. This is why we space the top bars in a beam as in a, as Fig 1.6 shows. Unit 2 (Example 2.1 on page 24) deals with this in more detail.

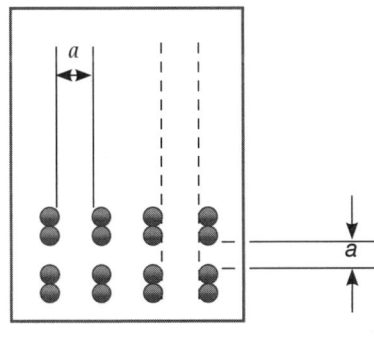

Fig 1.6 Spacing of reinforcement

To be able to use the specifications, you should study the extract from the South African Bureau of Standards SANS 282:2004 that can be found in the Appendix.

Only an extract from SANS 282:2004 has been included in the Appendix. You need to acquire and study the whole document.

1.3 General principles for drawing

Let us start talking about reinforced concrete structures by saying something about the sequence and manufacture of reinforced concrete structures. Such structures should start off with a base or foundation. We are only going to discuss cast-in-situ concrete that rests on simple spread foundations. Spread foundations may take the form of isolated bases (each base supporting one column) or strip foundations (each strip supporting a wall). These foundations are usually reinforced with steel bars to spread the concentrated load of the column or wall over a larger area of soil. The structural engineer determines the size of these foundations.

Remember that concrete is not as strong in tension as in compression.

Unless we reinforce bases, they are likely to fail in shear or bending. Fig 1.7 illustrates the nature of the failure that is likely to occur.

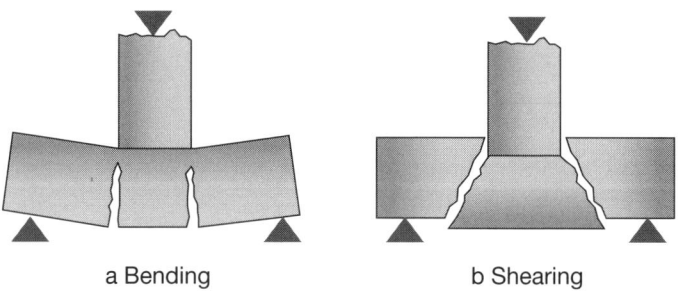

a Bending   b Shearing

Fig 1.7 Failure modes of a concrete pad foundation

The solution is to place the reinforcing steel where it will compensate for the low tensile strength of the concrete. (In the case of Fig 1.8 on page 9, the steel is placed at the bottom of the base.) Note that we have two layers of steel at the bottom of the base: the layer nearest the bottom of the base (B1) and the layer on top of this layer, which is perpendicular to the B1 direction (B2) and which ties down onto the first layer.

Module 1: Unit 1

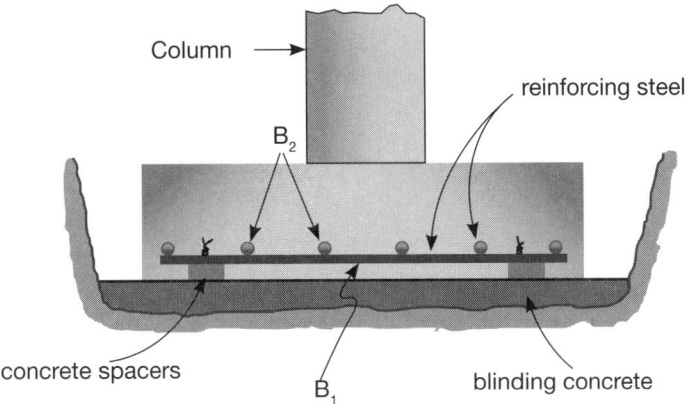

Fig 1.8 Reinforcement of a concrete base

We can cast the foundation in a hole in the ground where it will bear directly on the soil. However, this creates the problem of trying to keep the area clean while the reinforcing steel is being placed and the formwork fixed in position. To overcome this, a layer of blinding concrete (see Fig 1.8), about 50 to 75 mm thick, is usually laid below the designed base of the foundation. In addition, you have to protect the reinforcing steel against dampness from the soil by implementing a minimum thickness of concrete, known as the **concrete cover**. We accomplish this by the use of spacers or chairs, which support the steel at the required distance above the blinding concrete. Next, the columns are built on top of the bases. These will be reinforced to counteract bending, shearing and direct compression failures (see Fig 1.9).

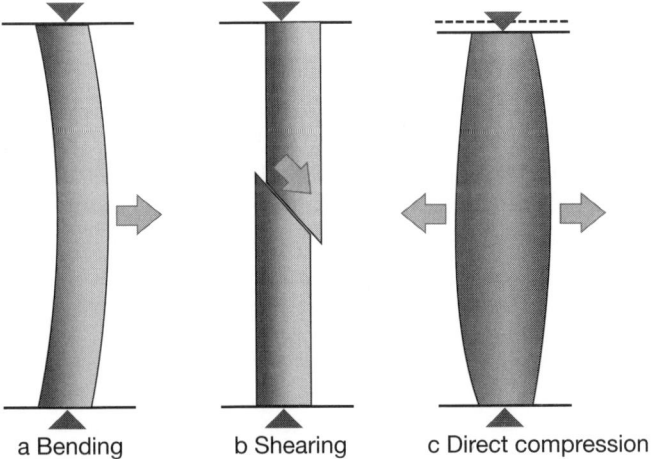

Fig 1.9 Failure modes of a concrete column

A typical column will have at least four longitudinal bars and a series of transverse bars, known as links. As illustrated in Fig 1.10, these links will compensate for the weakness of the concrete.

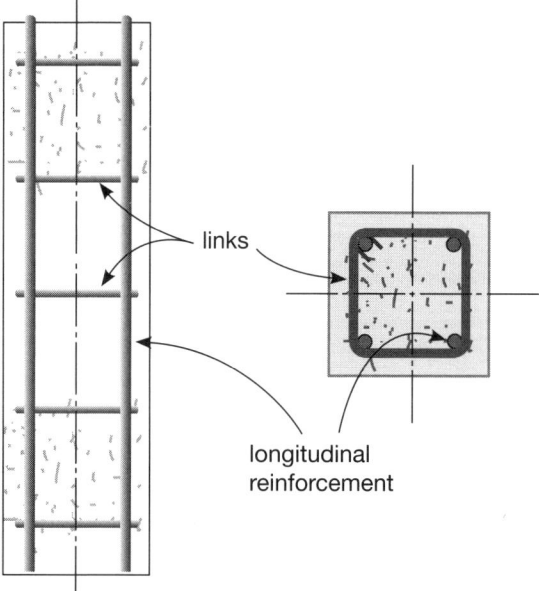

Fig 1.10 Reinforcement of a concrete column

The obvious question is: How do we combine the column with the base without causing a weakness in the structure? It is important to tie the column to the base with reinforcement. Consequently, starter bars are cast into the base. These starter bars project far enough above the surface of the base for the longitudinal column bars to be fixed to them by a process known as lapping. The definition sketches, Fig 1.11a and Fig 1.11b on page 11, should answer any questions that you might have. Note that the starter bars are bent through 90°, so that they can tie onto at least two bars of the foundation reinforcement, thus ensuring continuity from the base through to the column.

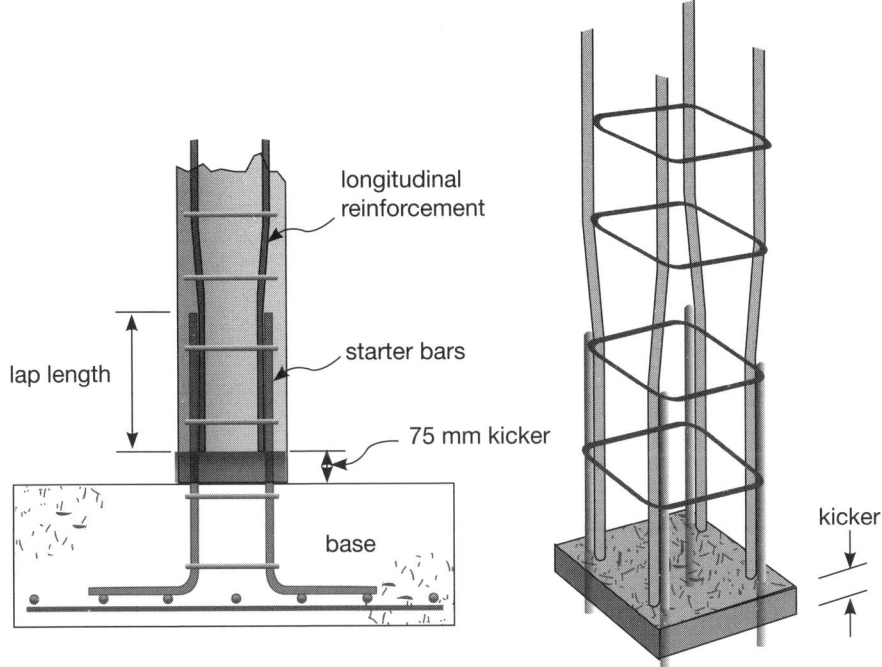

a. Junction of column with base   b. Isometric drawing of bars lapping

Fig 1.11

To form the column, we use formwork (shuttering). This formwork stands on the base. In order to locate the lower end of the formwork, it is usual to construct a very short length of column (about 75 mm), known as a kicker (see Fig 1.11b). The column is then concreted up to the underside of the first intersecting beam or floor slab. It is impossible to cast all the concrete in one day. It is therefore necessary to have joints to determine where each cast should start and where each cast should stop. The joints between the different days' pour of concrete are known as construction joints (CJ). There are standard places where these should occur, so as not to weaken the structure or make it difficult to build. The detailer should be aware of these positions, which are shown in Fig 1.12 on page 12.

The column bars have to be long enough to extend far enough above the top of the column to allow space for the depth of the beam and floor slab, the kicker for the next column and the necessary lap. The beam and floor slab can often be cast in a single operation, but it is common to cast larger beams up to the underside of the floor slab first, and then to cast the floor slab on another day.

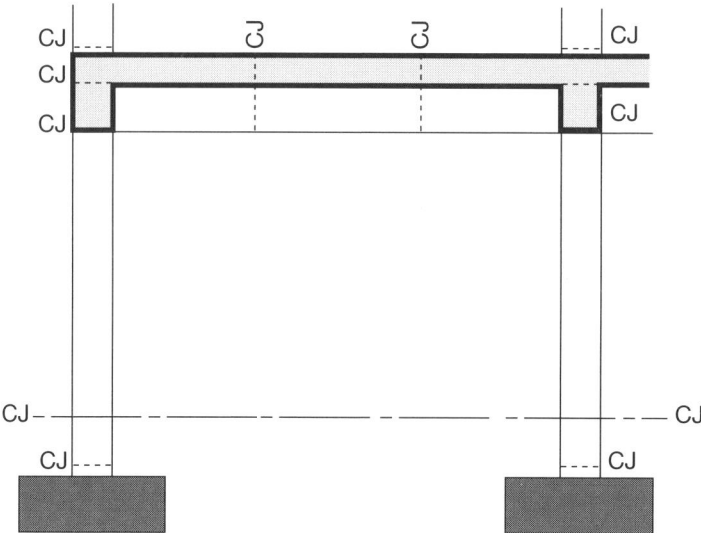

Fig 1.12 Siting of construction joints (CJ)

1.4 Bar schedules (or bending schedules)

A very important aspect that you need to take into account is the order in which the bars are fixed. On the drawings and schedules, the bars are numbered according to the order in which they are fixed. Table 1.2 on page 13 is a standard bar schedule or bending schedule, which we use to list the required bars for the reinforcement of concrete, using the standard shape codes as listed in the Appendix (see pages 240 to 247). A separate schedule is required and prepared for each structural element (member), and corresponds to the detail drawings. A filled-in bar and bending schedule is shown in Table 2.1 on page 30 in Example 2.1 (Unit 2). Bars that are required for multi-storey work are scheduled floor by floor. Schedules are prepared by the detailer and used by the builder, the reinforcement supplier, the steelfixer, the clerk of works and the quantity surveyor. It is therefore essential that each schedule should be a document complete within itself. We use the term 'bar schedule' because the fabricator will probably prepare separate cutting and bending lists that suit his purpose. While the bar schedule is prepared in the sequence of the structural elements (from foundations to roof), the cutting and bending lists are usually sorted into the type and size of the bar.

The schedule is completed as follows:
- **Column 1**

'Member' means the structural element for which the bars are scheduled, for example base, column and the like. It is customary to start with the foundations and progress through the building in the order in which it is likely to be constructed.

Table 1.2 Bar schedule

Member	Reinforcement					Bending dimensions					Mass	Fixing and non-standard bending details
Mark; Size; No. Off.	No. ea	Total no.	Size	Mark	Length (total length in mm)	Shape code	A	B	C	D	EorR	kg

'Number of members' means the number of similar elements that make up the structure.

■ Column 2

'Reinforcement' means the physical dimensions that the fabricator needs to know in order to cut the required reinforcing steel.

'Number of bars in each' means the number of similar bars of this mark in the member.

'Total number' means the product of the number of members and the number in each.

'Size' refers first to the type of steel:
- R = plain round mild steel bars of strength 250 MPa;
- Y = high yield deformed steel bars of strength 450 MPa; and
- Z = types of steel not covered by R or Y.

It then refers to the bar diameter in millimetres.

 Bar diameters are measured in millimetres (mm).

Bar 'mark' means the serial number allocated to the bar on the detail drawing. It is customary to list bars in mark order, which is the order in which they will be assembled in the formwork.

The 'length' of each bar in mm can be calculated only when the 'bending dimensions' have been completed. The length of each bar is specified to the nearest multiple of 50 mm.

■ Column 3

'Bending dimensions' (see Appendix, pages 240 to 247) provide these dimensions. Against each shape code there is a picture of the bent bar, with the crucial dimensions lettered A, B, C, D and E as required. R is the radius around which the bars of special shapes are bent.

■ Column 4

'Mass' means the total mass of the member for truck loading purposes. Use Table 1.3 on page 15 to find the mass per metre.

■ Column 5

'Fixing and non-standard bending details' refer to shape code 99, which should be shown and dimensioned if a standard shape does not exist. Examples 2.1 and 2.2 on pages 24 and 29 show bending schedules.

Table 1.3 Standard dimensions for rounds

Diameter (mm)	6	8	10	12	14	16	18	20	22	24
Mass (kg/m)	0,222	0,395	0,617	0,888	1,21	1,58	2,00	2,47	2,98	3,55
Diameter (mm)	25	27	30	32	35	40	45	50	55	60
Mass (kg/m)	3,85	4,49	5,55	6,31	7,55	9,86	12,5	15,4	18,7	22,2
Diameter (mm)	65	70	75	80	83	84	90	95	100	103
Mass (kg/m)	26,0	30,2	34,7	39,5	42,5	43,5	49,9	55,6	61,7	65,3
Diameter (mm)	108	115	120	130	134	140	145	150	160	165
Mass (kg/m)	71,9	81,5	88,7	104	111	121	130	139	158	168

Standard lengths: 6 m to 13 m in 1m increments

1.5 Types of drawings used for reinforced concrete

Now that we have discussed the requirements of detailing a reinforced concrete drawing, we can draw the reinforced concrete structure. Detailing reinforced concrete structures requires the production of two sets of drawings, that is, the concrete drawings and the reinforcing drawings. The structure is dimensioned on the concrete drawings and no reinforcing is shown. The detailing of the reinforcing is done on the reinforcing drawings and the dimensions of the structure are not shown. The line work in the two sets of drawings is different (as you will learn further on). Dimensions and detailing of the drawings are very important, and all sections and details must be clearly indicated.

1.6 Summary

1. Because concrete is strong in compression and weak in tension, and steel is strong in tension, we combine the two materials. We do this using design codes and tables and curtailment to design a structure that should be strong when subjected to axial forces, bending moments and, sometimes, torsion.
2. Now you know the sequence to follow when building reinforced structures such as bases, columns, beams and slabs.
3. In each of these structures you need to understand the meaning and use of elements such as bond and **lap lengths**, starter bars, blinding, links (stirrups), kickers and the like.
4. Now you know how to dimension the bars and how to complete the bending (bar) schedule.
5. You will use two types of detail drawings, namely concrete drawings and reinforcing drawings.

Self-evaluation

1. What do you understand by reinforced concrete?
2. Why is it necessary to reinforce concrete?
3. Explain what it means to detail reinforced concrete.
4. Explain the following description for dimensioning reinforcing bars: 22 R12-05-150T.
5. What sequence should you follow to manufacture a simple reinforced concrete structure?
6. What is a bar schedule?
7. Explain what is implied by steel type R and steel type Y.
8. Which types of drawings are required for reinforced concrete?

SELF-EVALUATION ANSWERS

1. Reinforced concrete is the combination of concrete and steel.
2. Concrete is not as strong as steel in tension and bending. Unless a concrete structure is reinforced by means of steel, it is likely to fail in shear and bending.
3. To detail reinforced concrete is to provide a schematic description of the way in which the steel is to be placed in a member.
4. Twenty-two mild steel bars of 12 mm diameter, bar mark 05 at 150 mm, centre-to-centre in the first top layer of the floor or slab.
5. Foundations/bases followed by columns, then beams, then floor slabs.
6. A bar schedule is a schedule that is used to list the required bars for the reinforcement of concrete.
7. Y steel is high yield deformed bars (450 MPa), while R steel is plain round bars (250 MPa).
8. Reinforced concrete requires concrete drawings (with no reinforcing shown) and reinforcing drawings (which detail the reinforcement).

Module 1: Unit 2

Unit 2 Foundations and columns

2.1 Introduction to foundations

We use reinforced concrete foundations, footings or bases to support columns and walls. Columns and walls can be made of concrete, steel, masonry and/or timber. Bases have to be capable of taking the total load that is distributed through the columns. The total load includes the weight of the roof, the floors, the beams, the columns and the live load. Remember that unless bases are reinforced, they are likely to fail in shear or bending.

2.2 Types of bases

There are two major types of bases, namely spread footings and pile footings, which can both be classified into sub-groups, such as isolated footings and combined footings.

2.2.1 Spread footings

We design spread footings to distribute large loads over a large area of soil near the ground surface, in order to reduce the intensity of the force per unit area so that the soil will safely support the structure.

2.2.2 Pile footings

We design pile footings to deliver large loads to individual 'piles'. The piles transfer the forces to lower levels within the soil by means of friction between the soil and the pile surface and/or end bearing. See Fig 2.1a on page 18.

Revise your *Construction* notes for clarity on these types of bases.

Isolated footings support the load of a single column. The foundation for a structure may be composed of several isolated footings and combined footings.

Combined footings support two or more columns, and act as beams or slabs resting on the soil or piles.

Other types of footings include the following:
- **Wall footings** usually support continuous concrete or masonry walls around the perimeter of a building. Interior partition walls may also rest on continuous wall footings.
- **Raft footings** may support many columns and walls, and act as continuous slabs to distribute the load over large areas.
- **Cantilever footings** may be used near property lines or other structures.

17

Fig 2.1b illustrates the types of spread footings that we use most frequently. These may be used with or without piles.

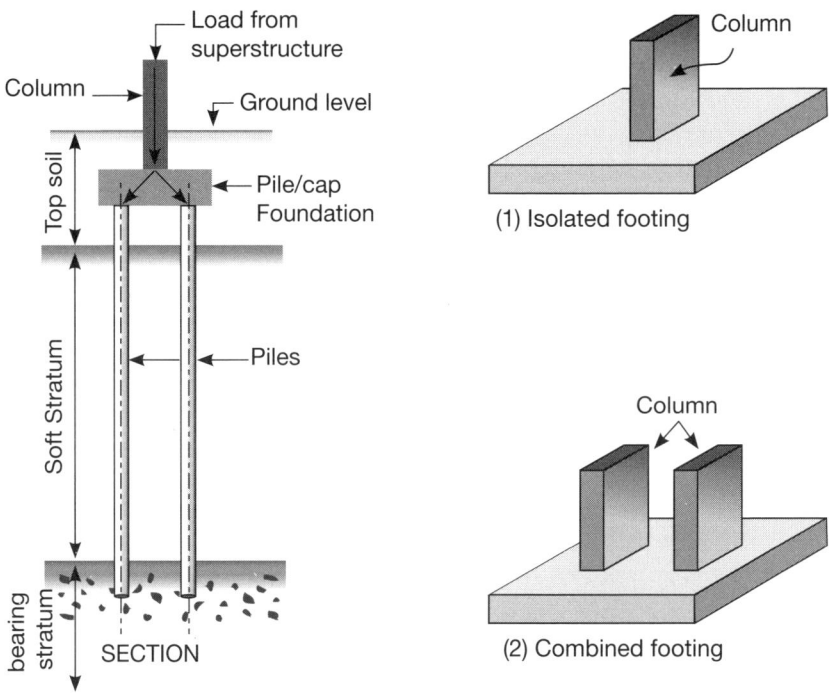

Fig 2.1a Pile footing

Fig 2.1b Types of spread footing

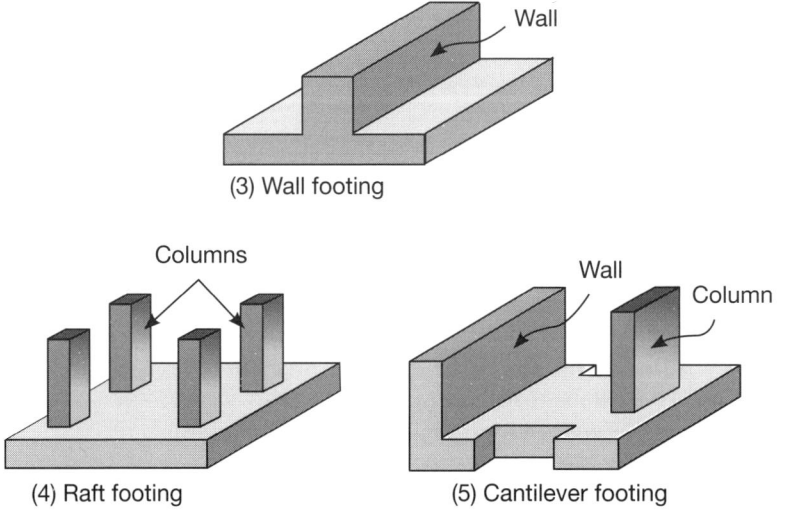

Fig 2.1b Types of spread footings (continued)

2.3 Introduction to columns

In cross-section, columns may be square, rectangular, circular or any other shape. Theoretically, columns could get smaller as the loading decreases towards the top of the building. In practice, however, we usually keep the columns the same size, and only reduce the quantity of reinforcement.

Before you start this section, turn back to Fig 1.11 on page 11.

2.4 Method of detailing columns

We detail columns in elevation with sufficient cross-sections to show the arrangement of the longitudinal bars and the shape of the links. A typical reinforced column is illustrated in Fig 2.2.

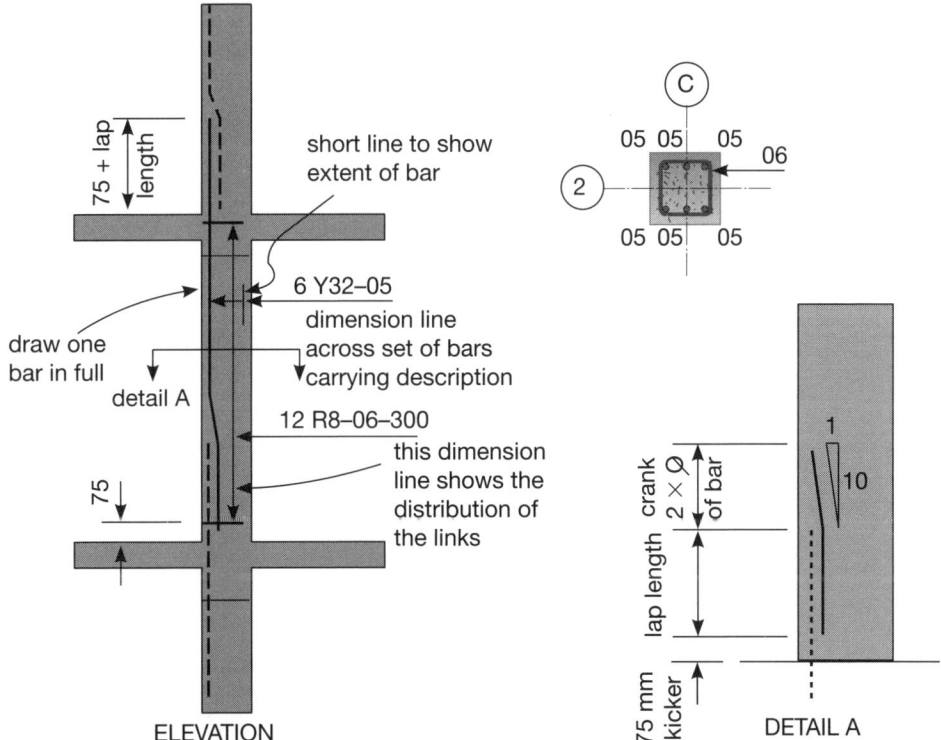

Fig 2.2 Reinforced concrete column

We show the longitudinal bars in elevation by one bar drawn in full and a short line to show the extent of the others. A dimension line is drawn across the set of bars, carrying their description; in this example six 32 mm diameter high yield bars (mark 05).

These bars stand on the kicker, and therefore start 75 mm above the foundation or floor slab. Because they cannot occupy the same space as the starter bars (indicated by a broken line), they have to be cranked as shown in the detail. The extent of the crank should be twice the diameter of the bar, rounded up to the next multiple of 5 mm. The slope of the crank should not be more than 1 in 10 (shape code 41).

The longitudinal bars are made long enough to act as starter bars for the next storey. Consequently, they finish by adding 75 mm (for the kicker) and the lap length above the next floor level.

We see the actual shape of the links in the horizontal cross-section. Their distribution is shown in the elevation by one link drawn in full and a short line to indicate the extent of the set of links.

Fig 2.3 illustrates some of the ways that we can arrange links to restrain the longitudinal bars in a column.

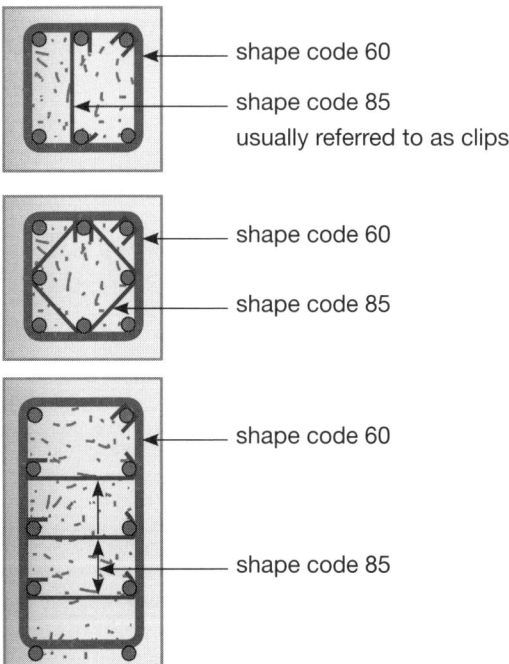

Fig 2.3 Some arrangements of links in columns

2.5   A closer look at footings and columns

In this course, we will only consider the simplest types of footings and columns.

2.5.1 Isolated footings

Let us consider Fig 2.4 on page 21 to understand the effect of the ground pressure on an isolated footing.

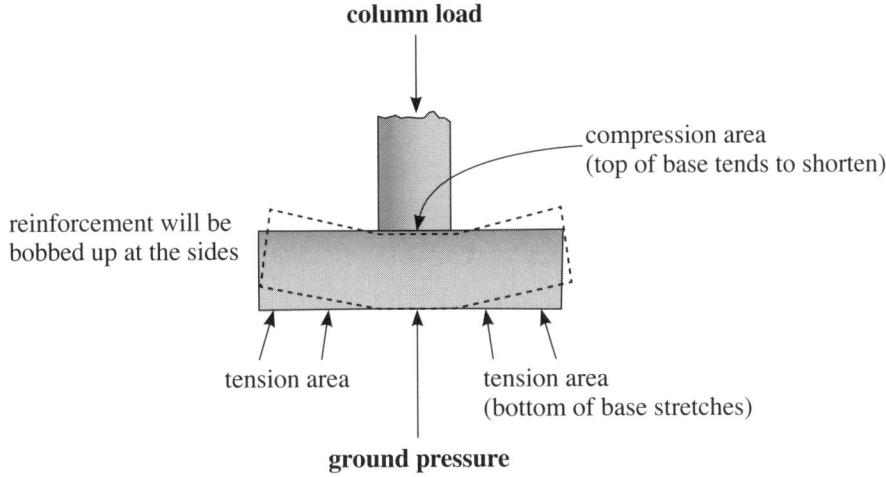

Fig 2.4 Effect of ground pressure on an isolated footing

As you can see, tension is caused in the underside of the base, so reinforcement is needed in the bottom only. It is a good idea to bob (bend) up the reinforcement at the sides to prevent layer cracking in the concrete (see Fig 2.8 on page 25). It is also good practice to use the same dimensions for the side concrete cover as for the underside.

 For more on compression lap, turn back to the definition of lap lengths on page 2 and Table 1.1 on page 3.

Starter bars, as described in Unit 1, are provided for the column and need to be long enough to provide sufficient compression lap above the kicker level. In this course, you are not expected to be able to design an isolated footing, so all the necessary information will be given to you. You need to ensure that you know how to apply the given information to produce a:
- concrete drawing;
- reinforcing drawing; and
- bending schedule.

2.5.2 Combined footings

Refer to Fig 2.5 on page 22 to understand the way in which the ground pressure will have an effect on this type of footing.

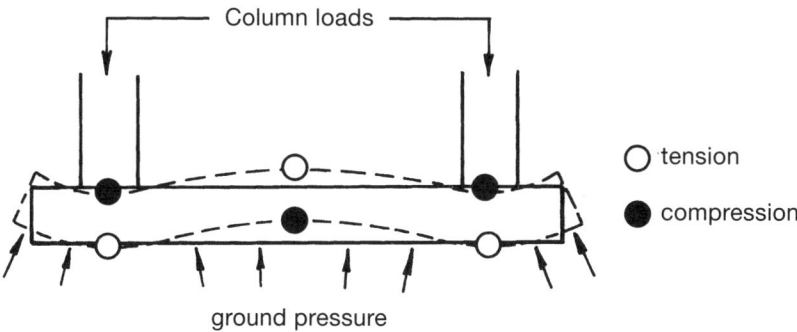

Fig 2.5a Combined footing

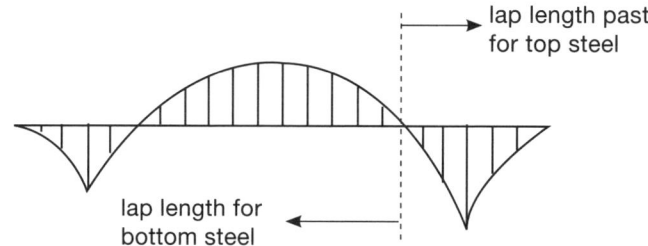

Fig 2.5b Combined footing bending moment diagram

Fig 2.6 on page 23 is a representation of the front elevation and plan of a reinforcing drawing. Refer to Fig 2.6 to help you understand the following explanation. This type of footing has tension in both faces along its length, due to the two columns causing a span moment in addition to cantilever moments. Consequently, reinforcement is required in the top between the columns and in the bottom outside the columns, as shown in the bending moment diagram. However, there should be some reinforcement in the bottom, throughout the length, so that in practice the lesser of the two steel areas required is carried through to provide a nominal lap with the other bottom steel. It also provides distribution (secondary) steel for the cantilever steel that is provided in the other direction.

Cross steel is provided in the top for distribution purposes and it helps to locate column starters, which are considered in the same way as with an isolated base. Cross steel in both the top and bottom of the base is bobbed and tied to provide support for the main top reinforcement. Note that the longitudinal steel at the top and the bottom, being the main reinforcement, passes outside the cross steel in order to form a greater lever arm.

The question that most definitely will arise is: How far past the supports should the top steel be completed? This question can only be answered when you, once again, consider the **45° dispersion angle** and find the bond length, as per Table 1.1 on page 3.

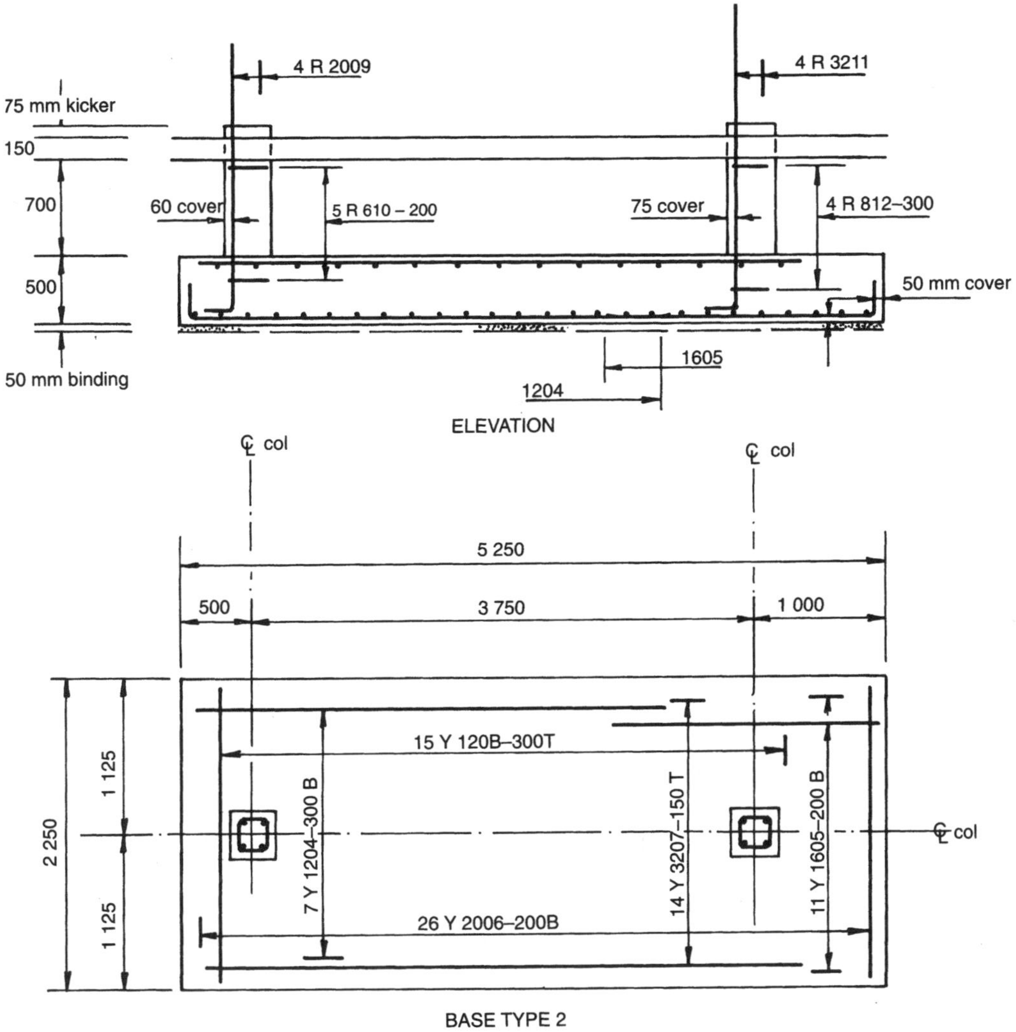

Module 1: Unit 2

Fig 2.6a Part of reinforcing drawing

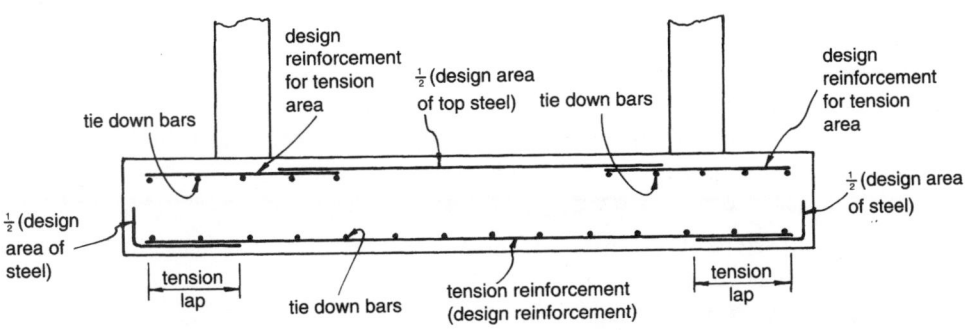

Fig 2.6b Part of reinforcing drawing

Drawing for Civil Engineering

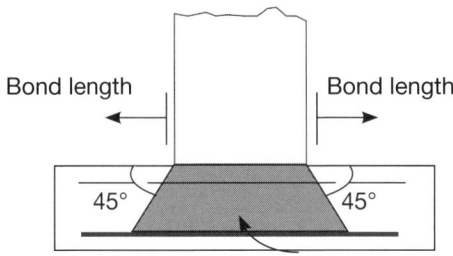

Fig 2.7 A 45° dispersion angle

2.6 Summary

There are different types of bases and each one has a different use. However, bases in general have only one use, and that is to support the total load right from the top of the roof, through the slabs and beams, through the columns onto the base, and not forgetting the ground pressure underneath the base. There are two types of bases, namely spread footings and pile footings — 'footings' being another word for bases or foundations. Furthermore, we classify these types of bases into sub-groups, such as isolated footings and combined footings. An isolated footing supports a single column, while a combined footing supports more than one column. Columns may be square, rectangular and circular, or any shape in cross-section, and are designed to resist the axial load and/or bending moment. The reinforcement consists mainly of four to a maximum of eight bars that run longitudinally in the column. The bars are supported by links to give the column a 'cage' form. Now that you have studied this unit, you should know how to use main steel and secondary steel, as well as tension steel and compression steel. This is very important, because it will determine the different bond lengths and lap lengths.

SOLVED EXAMPLES

EXAMPLE 2.1 Isolated footing

Use a scale of 1:20.

Draw the reinforcement layout and complete the bending schedule using the following information:
- Base size: 1 950 × 1 950 × 400;
- Column size: 280 × 280;
- Concrete cover to base: 50;
- Concrete cover to stirrups: 30;
- Concrete cube strength: 20 MPa (base and column);

- Base reinforcement: Y12 at 200 centres each way;
- Column starter bars: 4 Y12; and
- Links: R8 at 150 centres.

 Use a bar mark only once on the bending schedule.

Solution
(Read in conjunction with Fig 2.8.)

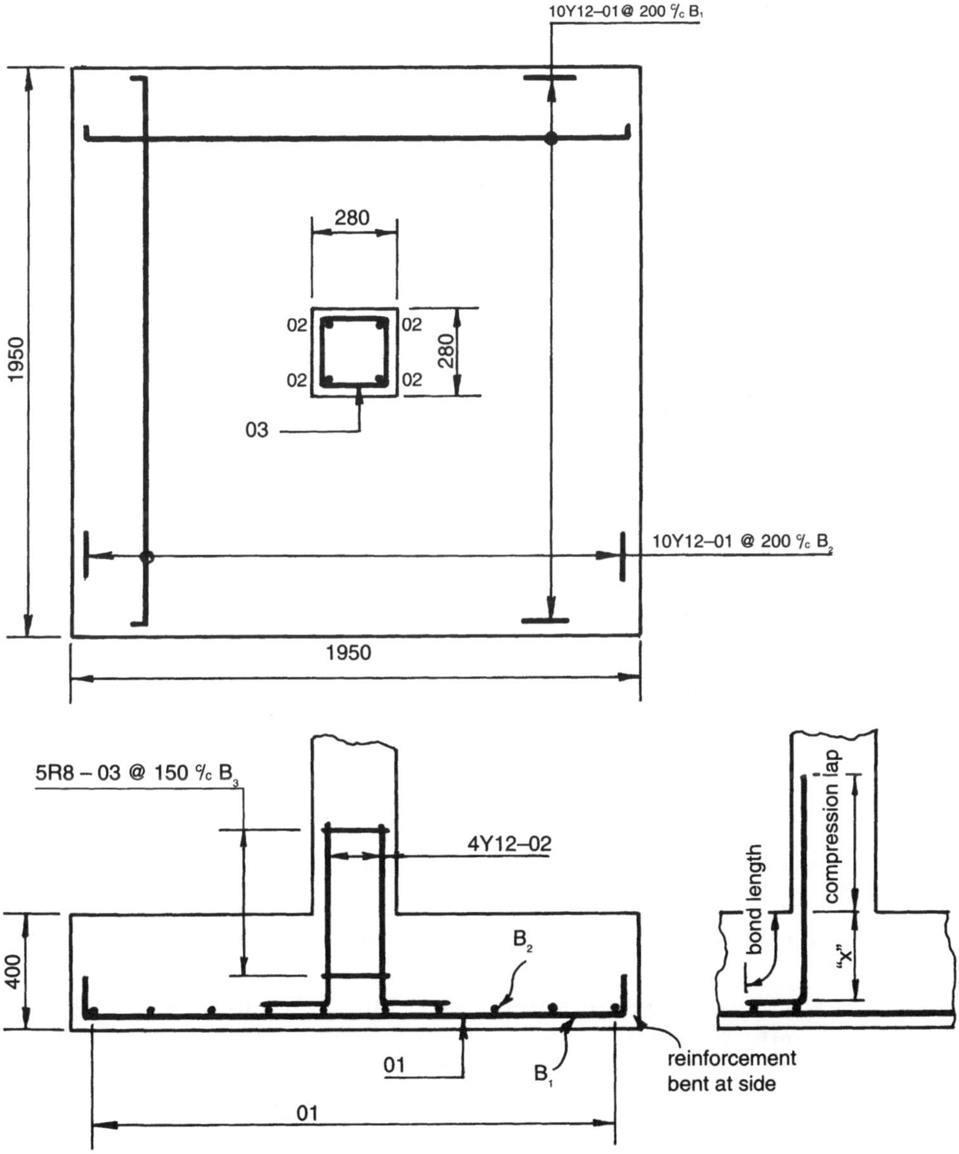

Fig 2.8

First, draw to scale the concrete outline (plan view and a section) of the base. On the plan, indicate the B_1 layer as shown; one bar is drawn in full, with two short lines to show the extent of the B_1 layer. We draw a dimension line across the set of bars, carrying their description. In reinforced concrete drawings, we call this the 'calling up of bars'.

Take note the little circle at the intersection of the one bar drawn in full and the dimension line. This circle indicates the bars that we are calling up.

Therefore in our B_1 layer we have 10 high tensile bars, each 12 mm in diameter, with a bar mark (reference) of 01. The bars are spaced at 200 mm, centre to centre, and are positioned at the bottom of the base, as with the first layer of steel. We calculate the number of bars as shown in Fig 2.9.

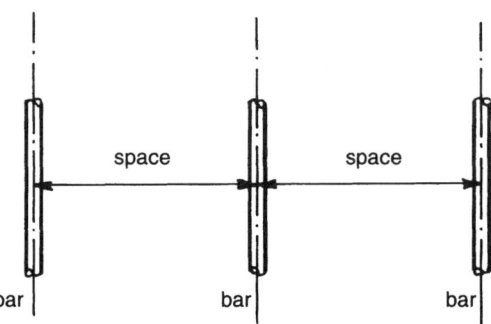

Fig 2.9 Isolated footing

From Fig 2.9 you can see that the number of bars = number of spaces + 1

$$= \frac{\text{width in span direction} - \text{concrete cover} - 2(\tfrac{1}{2}\text{ bar diameter})}{\text{pitch from c/c}} + 1$$

$$= \frac{1\,950 - 2(50) - 2(6)}{200} + 1$$

= 10,2 (Therefore, use 10 bars, that is, the spacing between bars will be slightly bigger than 200 mm.)

To determine the length of one bar, we use shape code 35. The cutting length is given by the formula: A + 2n (See SANS 282:2004.)
A = 1 950 − 2(50)
 = 1 850

Therefore, cutting length = 1 850 + 2(130) (*n* can be found in SANS 282:2004)
 = 2 110 mm
Remember to round up within 50 mm (see Definition 3.3 in SANS 282:2004), therefore length of bar = 2 150 mm

Total length = number of bases × number of bars × length per bar
 = 1 × 10 × 2 150
 = 21 500 mm

Mass = 0,888 kg/m × total length in metres = 0,888 × 21,5
= 19,092 kg (The mass in kg/m is read off the steel tables for round bars
— see Table 1.3 on page 15.)

On the section, indicate the B_1 layer as shown. Remember to call up the bar mark.

 Do you agree that in section we will see the true shape of only one bar?

Let's look at the B_2 layer. The B_2 layer ties down onto the first layer to prevent displacement of the B_1 layer, and is therefore placed perpendicular to the B_1 layer. You will notice that we use the same bar mark for this layer.

 Why is this?

This is done in accordance with the instructions in Example 2.1 to use a bar mark once only on a bending schedule. From our example we see that we have a square base, therefore the breadth is equal to the width. We are also using the same shape code; therefore the cutting length of the bars for the B_2 layer is equal to the cutting length of the bars for the B_1 layer. Therefore, when we call up the bars for the B_1 layer, the only change that we make is in the location of the bars, and instead of writing B_1 we write B_2.

On the section, indicate the B_2 layer as shown. Remember to call up the bar mark as shown in Fig 2.8 on page 25.

Let's look at the column starter bars (bar mark 02). In order to calculate the cutting length of these bars, it is easier to look at the section (since you will be able to see the true shape of the bar in section).

 Do you agree that in section we will see the diameter of all ten bars?

We use shape code 37 (see Appendix, pages 240 to 247).
Cutting length is given by the formula $A + B - \frac{1}{2}r - d$
$A + B - \frac{1}{2}(36) - 12$

(Where d = diameter of bar, and r = 36 SANS 282:2004)

To calculate dimension A, use Table 1.1 on page 3 to read off the bond length (for clarity on bond length, refer to detail on Fig 2.8 on page 25):
- Steel type (high yield deformed steel);
- Stress classification (tension);

- Concrete class (20 MPa); and
- Diameter of bar (12).

According to Table 1.1 on page 3, the minimum bond length = 530 mm.

Therefore, dimension A = bond length − dimension ×
$$= 530 − (\text{base depth} − \text{cover} − \text{diameter B}_1 − \text{diameter B}_2)$$
$$= 530 − (400 − 50 − 12 − 12)$$
$$= 204 \text{ mm (We use 210 mm, that is, round up to the next 10 mm.)}$$

If A is smaller than the B_2 centre-to-centre distance, we use the B_2 centre-to-centre distance instead of the calculated A distance. The reason for this is to make sure that the A dimension of the bar ties over at least two bars.

B dimension = x + compression lap
We use Table 1.1 on page 3 to read off the compression lap:
- Steel type (Y);
- Stress classification (compression);
- Column concrete strength (20 MPa);
- Bar diameter (12); and
- Compression lap = 390 mm.

Therefore, B dimension = x + compression lap
$$= 326 + 390$$
$$= 716 \text{ mm (We use 720 mm, that is, round up to the next 10 mm.)}$$
Therefore, cutting length = $A + B - \frac{1}{2}36 - 12$
$$= 210 + 720 − 18 − 12$$
$$= 900 \text{ mm}$$

On the section, indicate the column starter bars as shown. Remember to call up the bar mark as shown in Fig 2.8 on page 25.

We have to position our stirrups (R8) at 150 centres.

First, we have to calculate how many stirrups we need.

We normally position the first stirrup in the centre of the base, and the last stirrup approximately 20 mm from the top of the column starter bars.

Do you agree that in plan we will see the diameter of all four bars?

Module 1: Unit 2

This distance = compression lap − 20 + (base depth)
= 390 − 20 + 200
= 570 mm

So, number of stirrups = number of spaces + 1
$$= \left[\frac{570 - 2(\tfrac{1}{2}\text{ bar diameter})}{150}\right] + 1$$
= 4,72 (Therefore, we use five stirrups.)

Note the way we call up the stirrups in section.

 Do you agree that in plan we will see the true shape of the stirrup?

We use shape code 60 for the stirrup.
Cutting length = $2A + 2B + 2n - \frac{3n}{2} - 3d$ (For n, see SANS 282:2004)

Since we have a square column, the A dimension = B dimension
A = B = column size − 2(column cover)
= 280 − 2(30)
= 220 mm

Therefore, cutting length = $2(220) + 2(220) + 2(100) - \frac{3(100)}{2} - 3(8)$
= 906 (We make the cutting length = 950.)

With all our detailing and calculations completed, we must not forget to schedule our bars on the bending schedule. Your bending schedule should look like the one given on page 30.

Bar mark 01: Remember to add the 10 bars from the B_1 layer to the 10 bars from the B_2 layer (hence the 20 bars as scheduled). In addition, this changes the total length from 21 500 mm as calculated, to 43 000 mm as scheduled. Obviously, this influences the mass (kg) as well.

EXAMPLE 2.2 Combined footing

Use a scale of 1:20.

Draw the reinforcement layout and complete the bending schedule using the following information:
- Base size: 2 700 × 1 850 × 500;
- Column size: 500 × 250;
- Concrete cover to base: 60;
- Concrete cover to stirrups: 30; and
- Concrete cube strength: 25 MPa (base) and 30 MPa (column).

Table 2.1 Bending Schedule

Member	Reinforcement					Bending dimensions					Mass	Fixing and non-standard bending details
Mark: Size; No. Off.	No. ea	Total no.	Size	Mark	Length (total length in mm)	Shape code	A	B	C	D	EorR	kg	
Base 1 950 × 1 950 × 400 1 off	20	20	Y12	01	2 150 (43 000)	35	1 850					38.184	
Column 280 × 280 1 off	4	5	Y12	02	900 (3 600)	37	210	720				31.97	
	5	5	R8	03	950 (4 750)	60	220	220				1.88	

30

Module 1: Unit 2

Interpolate between 20 MPa and 30 MPa for your bond length.
Base reinforcement: Y20 at 200 centres Bl Y16 at 250 centres B2
Column starter bars: 6 Y12 (Use R8 clips — shape code 85)
Links — R8 at 150 centres

Use a bar mark only once on the bending schedule.

Solution

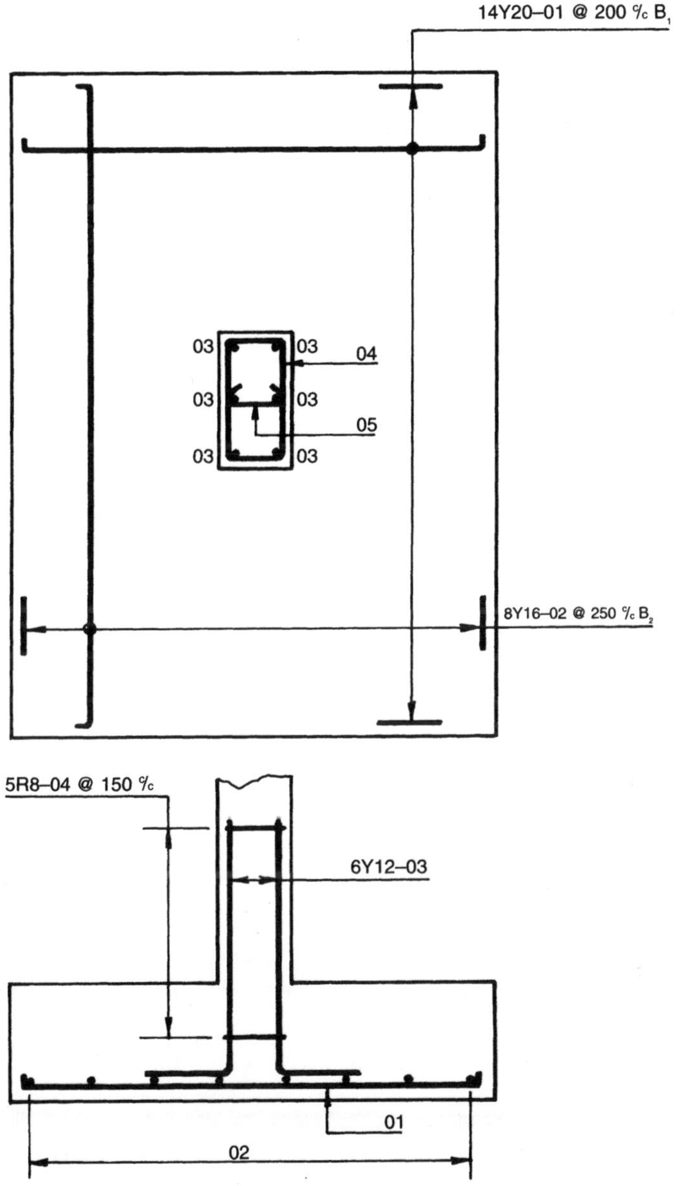

14Y20–01 @ 200 %c B$_1$

8Y16–02 @ 250 %c B$_2$

5R8–04 @ 150 %c

6Y12–03

01

02

Fig 2.10 Combined footing

31

The base in question has a rectangular shape. Therefore, the cutting length for the B_1 layer will differ from that of the B_2 layer; therefore, the bar marks will be different. The shorter bars normally become the main steel, and are positioned like the first layer, that is, at the bottom of the base, with the longer bars tying down perpendicularly onto them (that is, onto the first layer). With the detailed solution to Example 2.1 on page 24, you should be in a position to draw the reinforcement layout and produce a bending schedule. Let us explain the use of the R8 clips. As you can see from Fig 2.10 on page 31, the stirrups (shape code 60) hold the column starter bars in position. However, we include clips (shape code 85) to further stabilise the two-column starter bars positioned midway on either side of the longer side of the column. The number of clips is equal to the number of stirrups.

 Does this make sense to you?

The cutting length of the clips is given by the formula:
$A + B + 0,57C + D - \frac{1}{2}r - 2,57d$ where:
$C = 2$(diameter of clip) + diameter of column starter
$= 2(8) + 12$
$= 28$ (Therefore, make $A = 30$.)

Now, if you check SANS 282:2004, you will notice that C must be greater than or equal to $2r + 2r$.

Let's check whether this is in fact so: $2r + 2r = 2(16) + 2(16) = 64$

Note that our calculated C value is less than $2r + 2r$.
Therefore, we make our $C = 64$ (or 70 if we round off to the next 10 mm).
B = column width – 2(column cover)
$= 250 - 2(30)$
$= 190$

We can make the A dimension equal to the C dimension. We make the D dimension one-third the length of B:
$\frac{190}{3} = 63,3 = 70$ (rounded to the next 10 mm)

Therefore, the cutting length $= 70 + 190 + 0,57(70) + 70 - j(16) - 2,57(8)$
$= 341,34$ mm
(We use a cutting length of 350 mm.)

 Bar mark 03: A dimension = 250 because the calculated A dimension is equal to 66, which is less than the B_2 centre to centre distance.

Table 2.2 Bending schedule

Member	Reinforcement					Bending dimensions				Mass
Mark; Size; No. Off	No. ea.	Total no.	Size	Mark	Length (total length in mm)	Shape code	A	B	C	D	kg
Slab 5 000 × 2 540 × 170 (1 off)	14	14	Y20	01	2 100 (29 400)	35	1 730				72,62
	8	8	Y16	02	2 900 (23 200)	35	2 580				36,66
	6	6	Y12	03	1 050 (6 300)	37	250	800			5,59
	5	5	R8	04	1 300 (6 500)	60	440	190			2,57
	5	5	R8	05	350 (1 700)	85	70	190	70	70	0,69

To complete Example 2.2, you must:
- draw the complete concrete drawing;
- draw the side elevation of the reinforcing drawing; and
- complete a bar schedule for this combined footing.

Activity 2.1

(Answers not included)

The structure of a new building has 20 similar reinforced concrete columns and bases.

The size of a typical base is 1 900 × 1 900 × 450 mm thick, and the size of a typical column is 300 × 300 × 3 000 mm long.

The column is centrally positioned on the base.

Blinding concrete 75 mm thick is specified below each base because of the wet conditions.

Reinforcing in the base consists of Y12 bars at 125 centres to centres. These bars have a right-angle bend at each end and the concrete cover to these bars is 75 mm.

There are 4 × Y16 column starters and 4 × Y16 bars in each column. The links are R10 bars at 200 mm centres. The concrete cover (to the links) is 40 mm. Use a 75 mm kicker. Use Table 1.1 on page 3 to determine the bond and lap lengths.

Note that in our examples we did not make use of a kicker. However, in this activity your bond length will include the kicker.

The concrete cube strength is 30 MPa. Use a scale of 1:20.

Prepare for a typical column and base:
- fully dimensioned concrete drawings;
- separate, fully detailed, reinforcing drawings; and
- a bending schedule for the total steel to be ordered for all the columns and bases on the job.

The solutions to this activity are not provided. Ask your lecturer to check your drawings and schedule.

Activity 2.2

(Answers not included)

Prepare a fully dimensioned concrete working drawing for a reinforced concrete column and base to the following specification:
- Blinding concrete layer is 75 mm thick;
- Base size: 1 800 × 2 000 × 500 mm deep;
- Column size: 300 × 500 × 3 000 mm long;
- The column is centrally positioned on the base and the longer side of the column is parallel to the longer side of the base; and
- Use a 50 mm kicker.

Prepare a fully described reinforcing drawing for the above column and base, showing all the steel reinforcing to the following specification:
- Base steel: Secondary steel Y10 at 150 c/c;
- Main steel Y12 at 200 c/c;
- These bars all have right-angle bends at each end, as in Fig 2.11;

Fig 2.11 Right-angle bends
- Column starter: 6 × Y16;
- Column steel: 6 × Y16;
- Links (stirrups): R8 at 200 c/c. Provide R8 clips as well;
- Concrete cover: Base 75 mm; and
- Column 40 mm.

Prepare a bending schedule for five columns and five bases.

Concrete cube strength is 30 MPa. All dimensions are in mm. Use a scale of 1:20. The solutions to this activity are not provided. Ask your lecturer to check your drawings and schedule.

Module 1: Unit 2

Self-evaluation

1. What is the function of:
   a. footings; and
   b. columns?
2. When considering a foundation and the ground pressure which it experiences, where would you provide tension steel?
3. What type of steel would normally be provided in a column?
4. Name the types of footings that are generally in use.
5. What do you understand by:
   a. a kicker;
   b. a starter bar; and
   c. links?

SELF-EVALUATION ANSWERS
1. a. Footings have to carry the vertical load that is generally being transmitted through a column onto a base.
   b. Columns have to carry the load from the roof, through the slabs and beams as well as the loads of the columns above the footing.
2. You would provide tension steel at the bottom of a base.
3. Compression steel would normally be provided in a column.
4. Main types of footings: spread; pile
   Other types of footings: isolated; combined; wall; raft; cantilever
5. a. In order to locate the lower end of the formwork, it is usual to construct a very short length of column (about 75 mm), known as a kicker (see Fig 1.11b on page 11).
   b. Since it is necessary to tie the column to the base with reinforcement, starter bars should project far enough above the surface for the longitudinal column bars to be fixed to them by a process known as lapping (see Fig 1.10 on page 10).
   c. Links are a series of transverse bars which compensate for the weakness of the concrete (see Fig 1.10 on page 10).

Unit 3 Beams and slabs

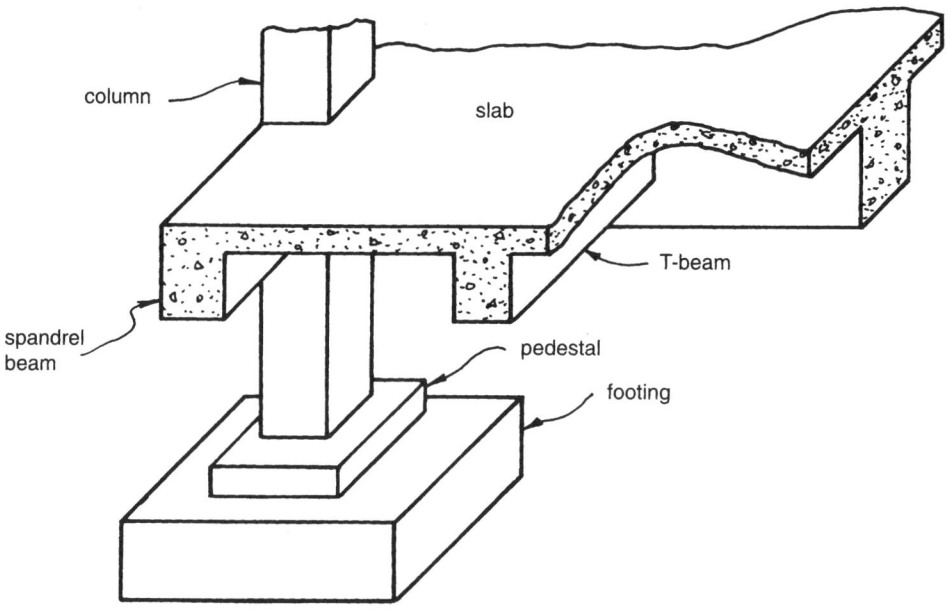

Fig 3.1 Beam and slab construction

3.1 Introduction to beams

In Unit 1, beams were mentioned briefly to show how they appear in the sequence of a building structure, but now (and later in Module 2) we will look at beams more closely.

Beams are structural elements, designed to carry external loads. Beams experience bending moments, shear forces and torsion moments along their length. These were all discussed in *Theory of Structures*. Beams can be simply supported (as Fig 3.2 on page 37 shows), fixed or continuous.

The structural engineer will consider bending, tension, shear and torsion, and will check whether the beam is able to transmit forces, especially where steel needs to be lapped, without causing internal cracking.

The shape of a beam can be square, rectangular, flanged or tee. Where beams are carried over a series of supports, they are called continuous beams, as is shown in Fig 3.3 on page 38.

Module 1: Unit 3

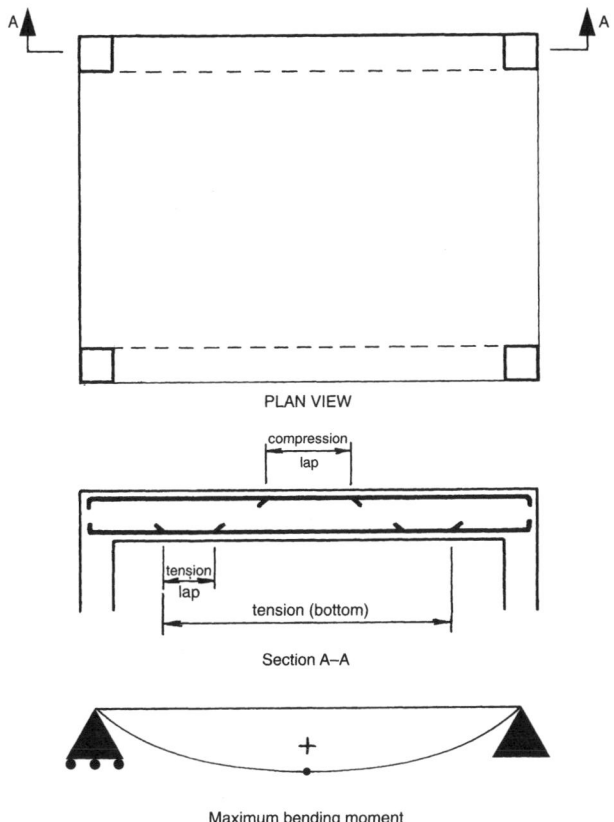

Fig 3.2a Simply-supported beam: plan view, section and bending moment diagram

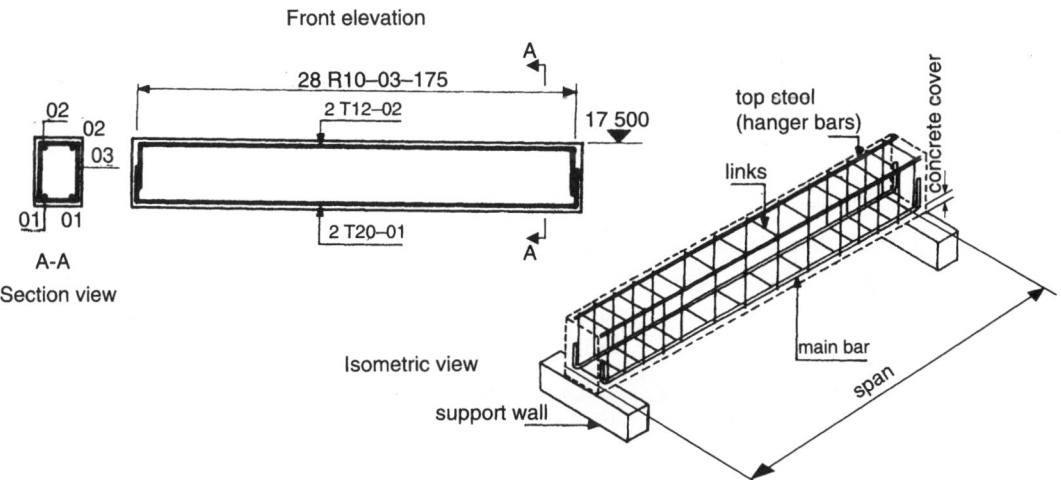

Fig 3.2b Simply-supported rectangular beam: front elevation, section view and isometric view

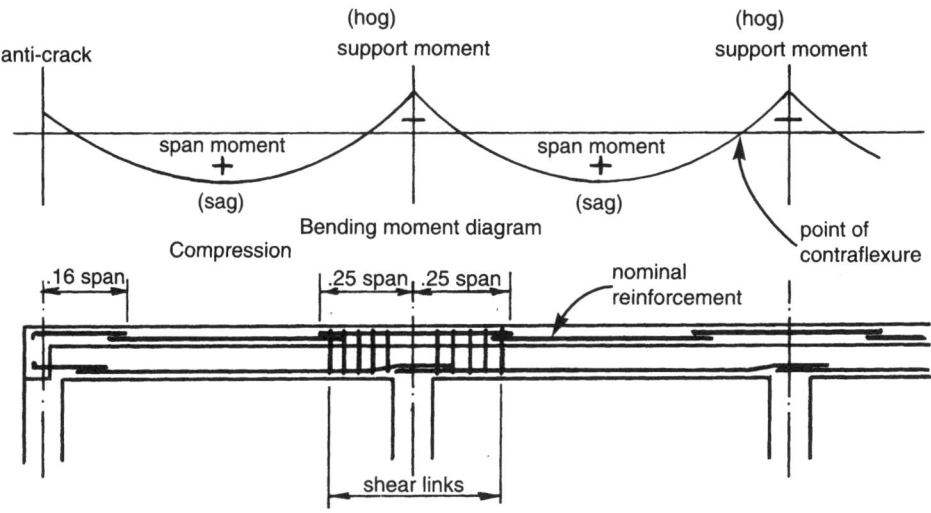

Fig 3.3 Continuous beam

Let us explain the bending moment of a continuous beam. This should sound familiar because you have calculated the point of contra flexure in *Theory of Structures*. A simple beam bends under a load, and a maximum positive bending moment exists at the centre of the beam (see Fig 3.2b on page 37). Tension therefore occurs in the bottom of a beam. In continuous beams, the sag at the centre of the beam is coupled with the hog at the internal support, resulting in a negative bending moment at the support. Where a positive moment changes to a negative moment, a point of contra flexure or inflection occurs, at which the bending moment is zero.

 This should sound familiar because you have calculated the point of contra flexure in *Theory of Structures*.

3.2 Detailing beams

A simple type of beam is usually rectangular in cross-section. It is a good idea to make the beams' width the same as the width of the columns that support them. The depth of beams usually includes the thickness of the floor slab. Owing to the nature of its bending, the most simple beam has tensile reinforcement in the bottom, as well as a link or stirrup reinforcement to resist shearing forces across the depth of the beam, as is shown in Fig 3.4 on page 39.

Two smaller-diameter bars are provided in the top of a beam, even if they are not required for strength. If you study a stirrup or link (shape code 72), you need at least two bars to 'hang' the stirrups onto. Hanger

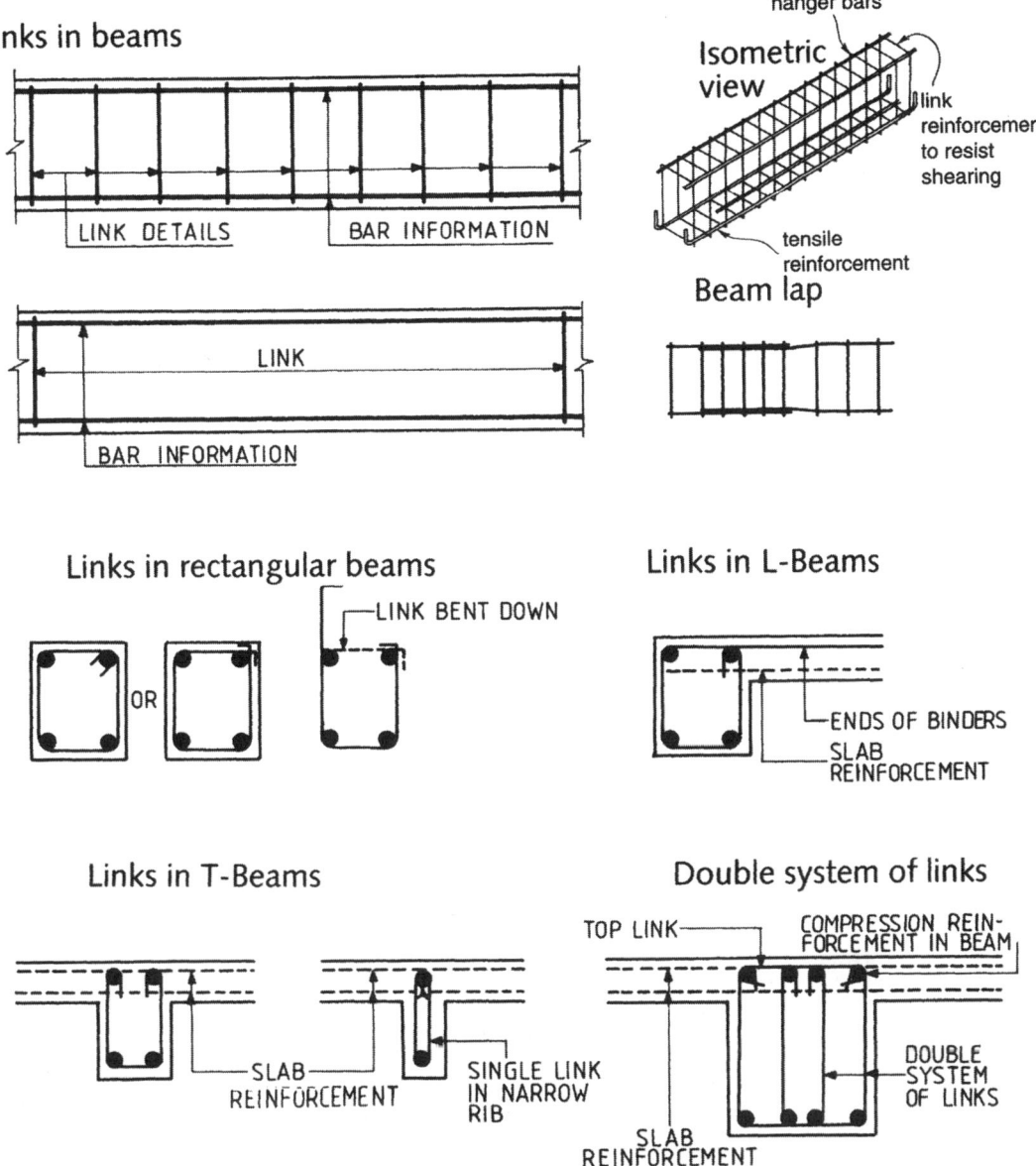

Fig 3.4 Links and bent bars

bars act as nominal reinforcement, and they enable the formation of a rigid cage, which will not collapse when the concrete is placed. In most cases, beams extend over several spans and, as previously mentioned, we call these 'continuous beams'. We have also mentioned that in such beams, in addition to the bottom areas between supports, the tensile zone also occurs at the top over the internal supports (see Fig 3.5 on page 40).

Drawing for Civil Engineering

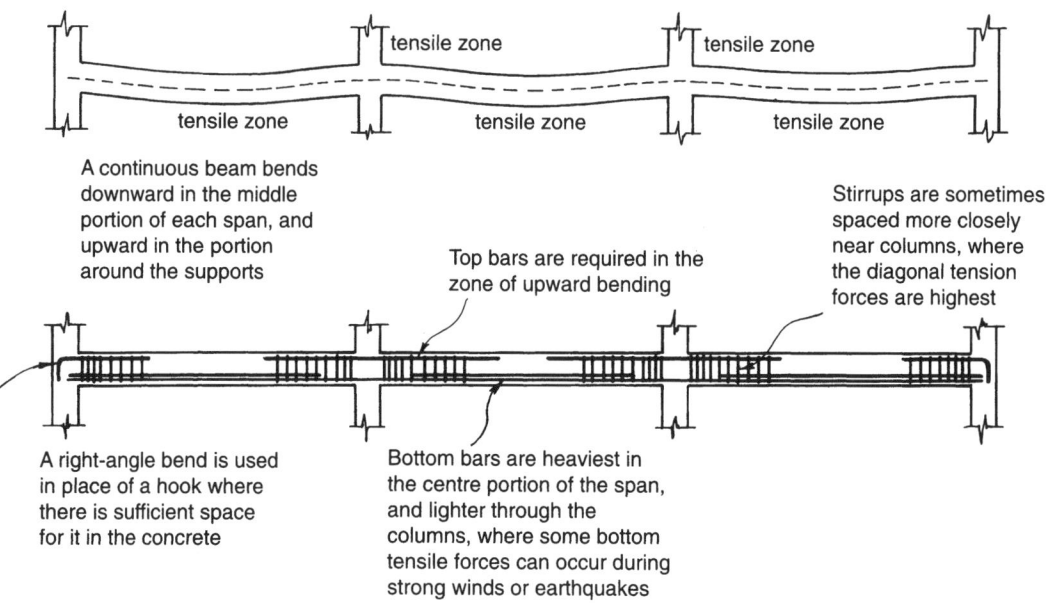

Fig 3.5 Reinforcing for a continuous beam

Fig 3.5 shows reinforcing for a concrete beam that is continuous across several spans. The upper diagram shows, in exaggerated form, the shape taken by a continuous beam under a uniform loading; the broken line is the centre line of the beam. The lower diagram shows the arrangement of bottom steel, top steel and stirrups conventionally used in this beam. The bottom bars are usually placed at the same level, but they are shown on two levels in Fig 3.5 to demonstrate the way in which some of the bottom steel is discontinued in the zones near the columns. Beams are detailed in elevation, with sufficient cross-sections to illustrate the positions of all the longitudinal bars and the shape of the stirrups or links. All descriptions of bars are given on the elevation, and only the bar marks are repeated in the cross-section.

Fig 3.6 on page 41 shows the detail of a typical beam that is continuous over two bays. You will see that no attempt is made to draw the elevation to scale, because this would produce a long, thin elevation. Therefore, we exaggerate the depth enough to make the detail clear, without making it look ridiculous. For the same reason, the cross-sections are also drawn to a larger scale. Although there are several bars in the top and the bottom of a beam, only one line is drawn

Module 1: Unit 3

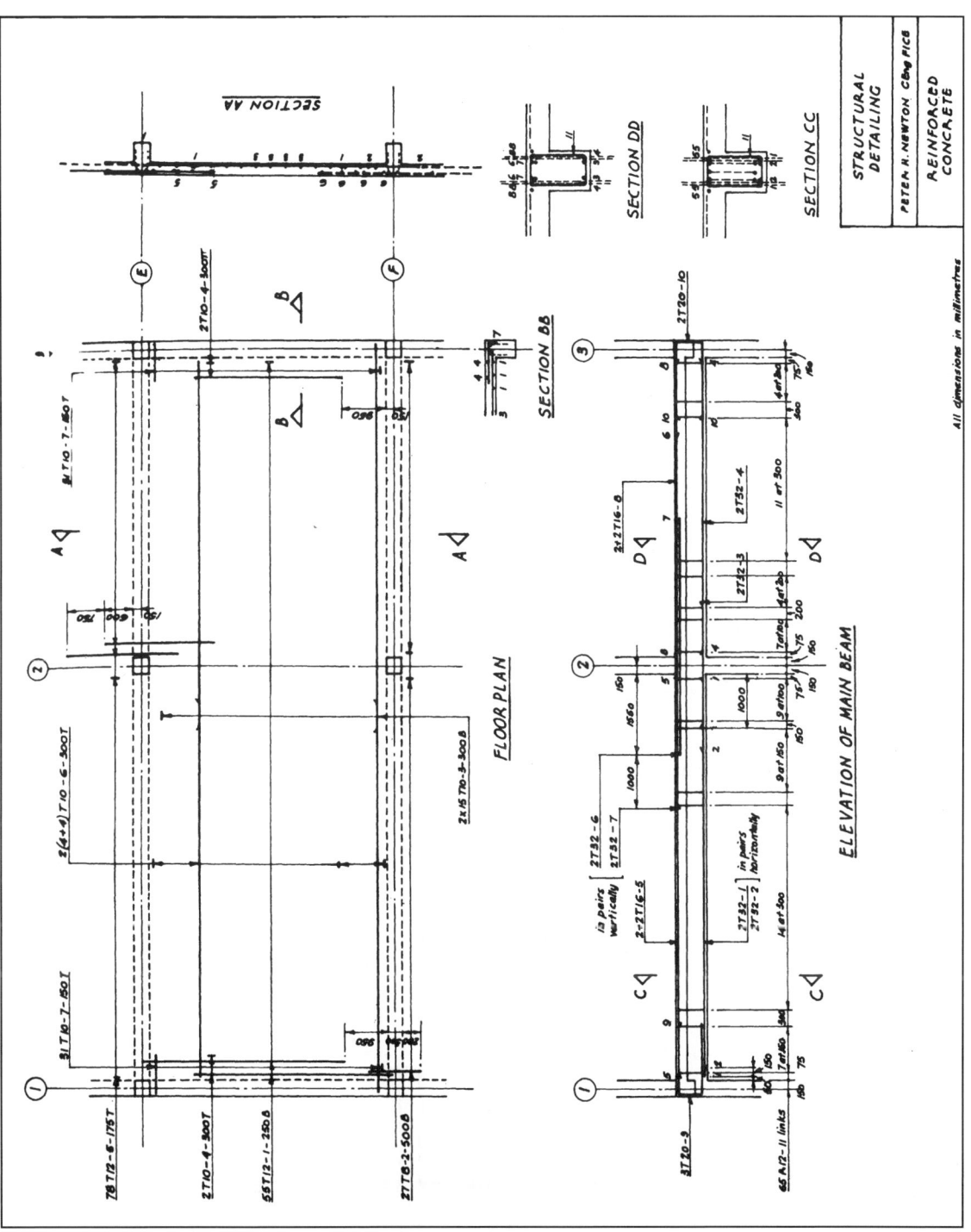

Fig 3.6 Typical floor and beam detail

41

to represent them. If a bar does not run the full length, ticks indicate its extremities, with the bar mark written alongside. These ticks do not mean that the bar is bent. Fig 3.7 shows this.

Fig 3.7 Ticks indicating bar extremities

Let us consider the supports of a beam. Over the support there are two layers of bars in the top of a beam, to cater for the high tensile forces in that area. The reason for this is also to leave a gap in the middle of the beam to insert a poker vibrator to compact the concrete. The bars, which are marked 7, may be in contact with the bars marked 6 — a vertical pair in this case. A vertical gap has to be maintained between two sets of bars, so spacer bars are used, as shown in Fig 3.8. These spacer bars must be scheduled.

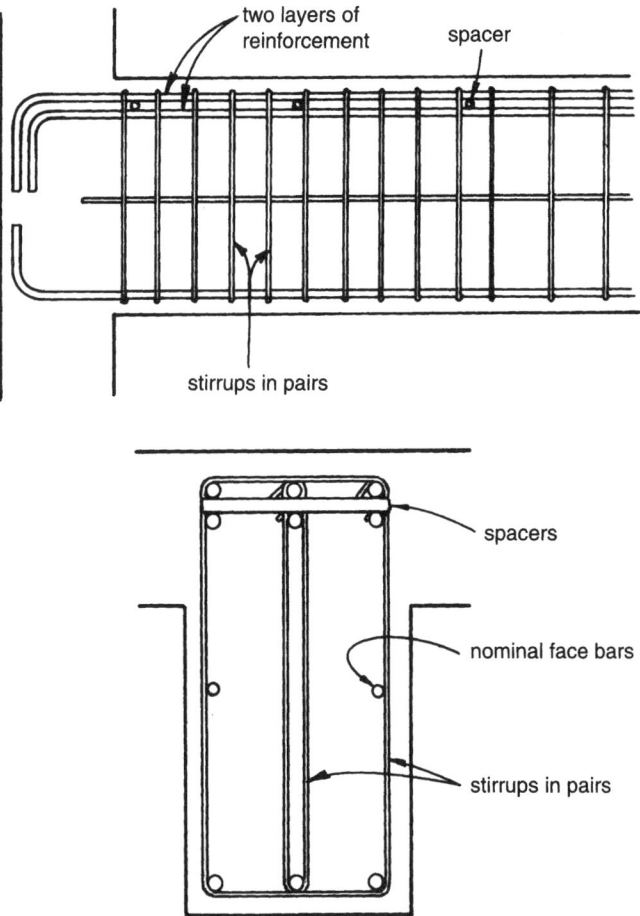

Fig 3.8 Provision of spacer bars

Another case to consider is one in which there are many bars in a beam. The problem is that one stirrup may not be enough, and therefore multiple stirrups should be fixed in sets. Fig 3.9 illustrates how stirrups are commonly arranged in beams.

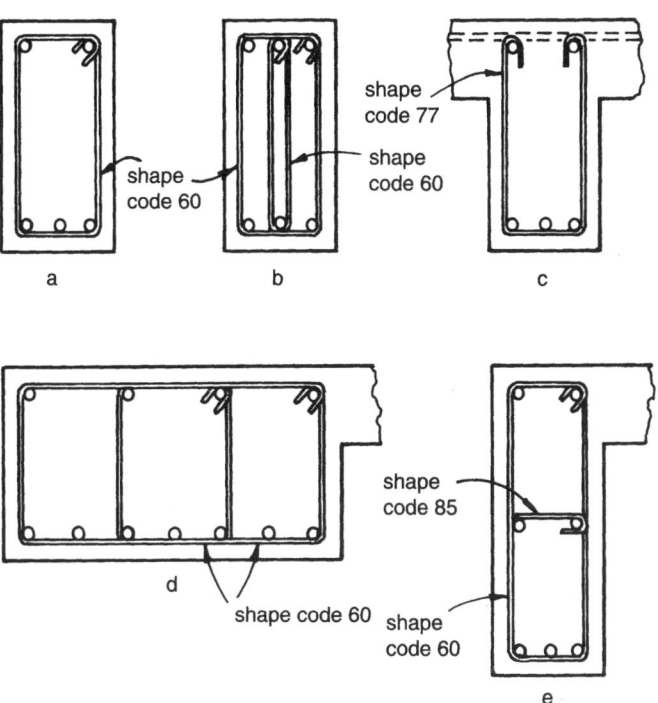

Fig 3.9 Some arrangements of stirrups in beams

Forms (a) and (b) show normal closed stirrups, (b) being used if the top of the beam is in compression, to provide lateral restraint to all the top bars. Form (c) shows open stirrups, because the floor slab reinforcement will provide the closing steel. Form (d) is a wide beam in which sufficient stirrups are used to ensure the rigidity of the cage. Form (e) is a deep beam where nominal, face reinforcement of the sides of the beam stiffens the stirrups when the concrete is placed and prevents cracking of the beam sides.

We describe stirrups in full in the elevation, and the section shows the shape of the stirrups.

The shape and size of the stirrups may be the same, but their spacing may vary to suit the change in magnitude of the shearing forces along the beam. These are greater nearer the supports, so the stirrups are close together there. The system of notation shown in Fig 3.8 on page 42 is recommended. Sometimes stirrups alone cannot counteract the shearing forces in the beam. In such cases bent-up bars are used, as shown in Fig 3.10 on page 44.

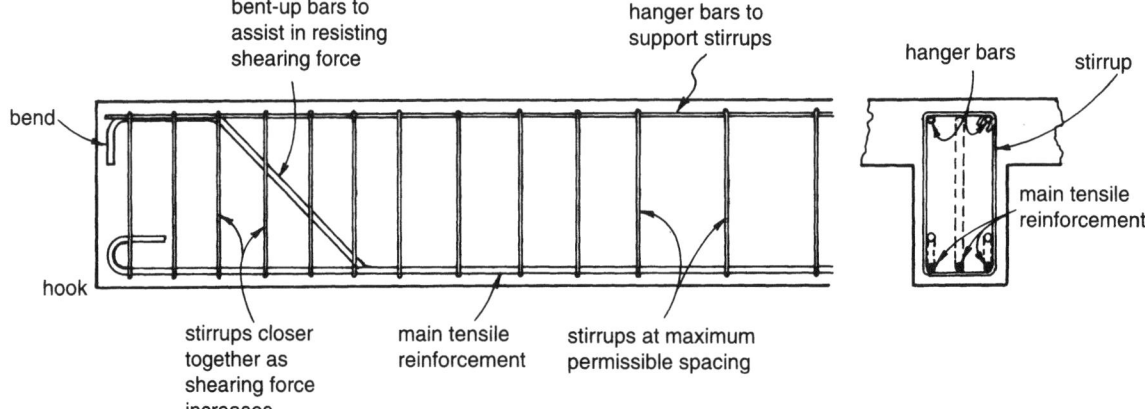

Fig 3.10 Use of bent-up bars to counteract heavy shearing forces

Instead of terminating bars that are no longer required for tensile purposes, the bars can be bent up (or down) through an angle of 45° so that they pass through the area where the shearing forces are at their greatest, and in so doing help to resist the forces. (The shearing forces are at their greatest where there is a change from tension to compression.)

3.3 Introduction to slabs

Slabs are divided into two categories, namely suspended slabs and supported slabs. Suspended slabs may further be divided into two groups. Suspended slabs can either be supported on edges of beams and walls, or they can be supported directly on columns without beams (which are known as flat slabs).

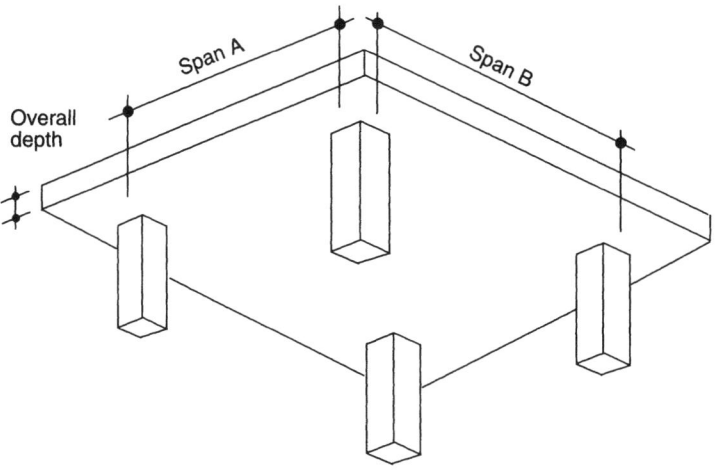

Fig 3.11 Typical flat slab

Supported slabs may be:

■ **One-way slabs**

These are slabs supported on two sides, with main reinforcement in one direction only. Fig 3.12 shows that the main reinforcement is provided along the shorter span. In order to distribute the load, distribution steel is necessary, and it is placed on the longer side. For clarification, see Fig 3.13.

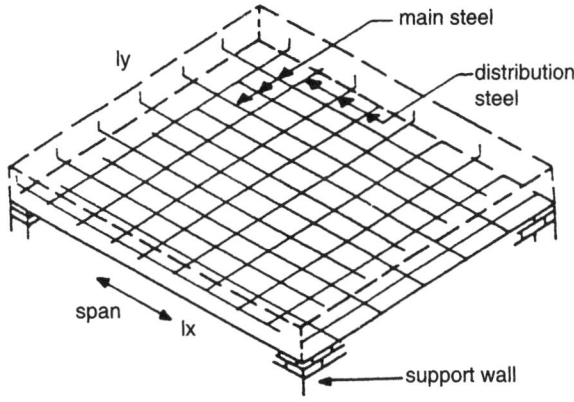

Fig 3.12 Slab reinforcement

ISOMETRIC VIEW

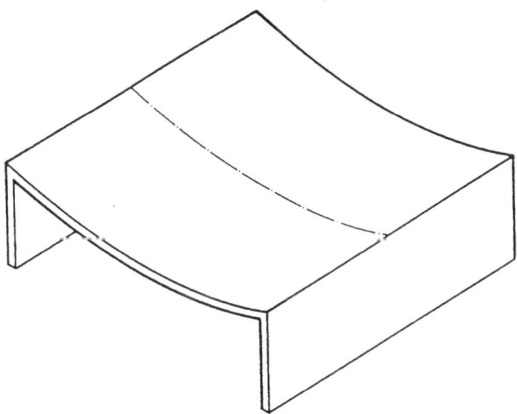

Fig 3.13 One-way slab action

■ **Two-way slabs**

These are slabs supported on four sides and reinforced in two directions. The principal values of the bending moments determine the size and number of reinforcement bars in each direction. A typical layout for a simply-supported two-way slab is shown in Fig 3.14 on page 46.

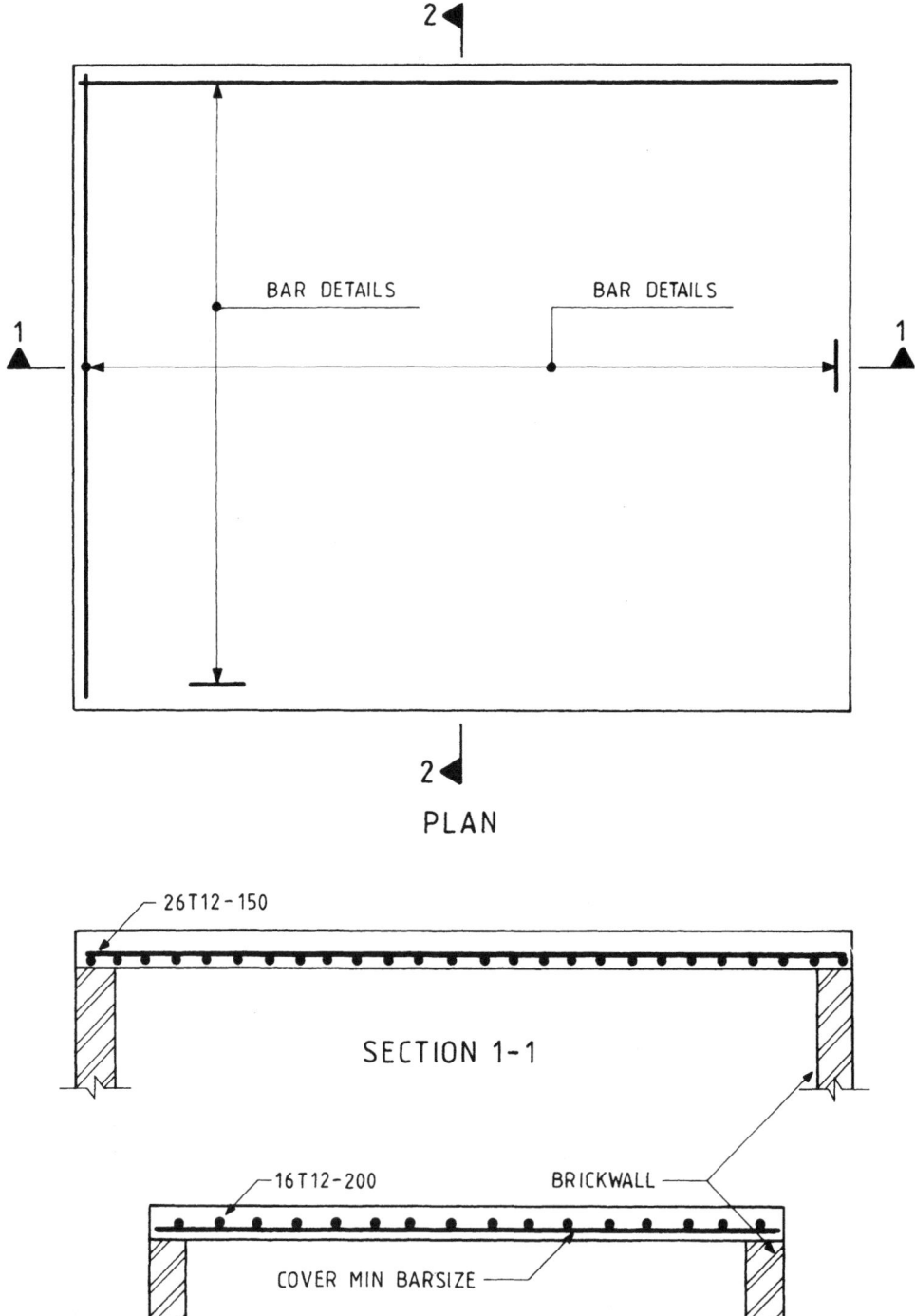

Fig 3.14 Simply-supported two-way slab

3.4 Detailing floor slabs

Floor slabs are designed differently from beams. The simplest floor slab spans in one direction, and is supported by beams or walls along opposite sides. Alternatively, floor slabs can be made to span in two directions at right angles, and are supported by beams on all four sides. As mentioned previously, there is also a type of floor that has no beams, but which acts as a plate that is supported on columns alone. Only in very small structures, such as garages, are slabs simply supported by just resting on supports. In most constructions the floor is rigidly held at the sides, or it is extended over several supports and continuous beams. Similar to a beam, the tensile zone is at the bottom of the slab in the middle of the span, and at the top over the supports.

Floor slabs are detailed in plan, with sufficient sections to show the positioning of all reinforcement. Descriptions of bars are given in full on the plan view, and only the bar marks are repeated in the sections. If possible, dimensions to show the positions of the ends of bars are given in the section, rather than on the plan. Sets of similar bars should be indicated by one bar drawn in full, with short lines to mark the extreme bars with a dimension line across the set of bars carrying their description. Where bars are staggered, or when alternate bars are reversed, it is common practice to show a pair of staggered or alternate bars in full on the plan.

No attempt should be made to indicate the bent shape of the bars on the plan. This can be done in the sections. In thin slabs, care must be taken to ensure that standard hooks and bends can be accommodated without reducing the cover. Secondary or distribution steel is always provided in slabs. This spreads any point load sideways over the primary reinforcement to form a rigid mat, which prevents bars from being displaced by the wet concrete. The lower layer of reinforcement is supported on spacers of thickness appropriate to the cover required, but the upper layer requires chairs to support it. Chairs may be formed out of reinforcement (using code 99) to support the upper layer. Code 99 requires a dimensional sketch to be drawn over columns A to E of the bending schedule. Careful consideration must be given to the chairs (see Table 3.1 on page 48), as they frequently carry not only the weight of the upper steel, but also the weight of men and equipment. Where slabs are extensive, timber screeds are sometimes fixed to the top steel so as to determine the top level of the concrete.

Table 3.1 Typical chairs

Bar support ilustration	Bar support illustration Plastic capped or dipped	Type of support
	capped	Slab bolster
		Slab bolster upper
	capped	Beam bolster
		Beam bolster upper
	dipped	Individual bar chair
	dipped	Joist chair
	capped	Individual high chair
		High chair for metal deck
	capped	Continuous high chair
		Continuous high chair upper
		Continuous high chair for metaldeck
	dipped	Joist chair upper

Beam reinforcement is shown in broken lines in the section to ensure that consideration is given to its position relative to the slab steel. The upper reinforcement in the slab usually passes over the beam steel, and in so doing may reduce the concrete cover, unless the beam steel is kept low enough. If the slab is heavily reinforced, leading to a complicated drawing, the upper and lower layers may be shown on separate plan views. Refer to Fig 3.15 for more on reinforcing for a one-way concrete slab. The reinforcing is similar to that of a continuous beam, except that stirrups are not usually required in the slab, and shrinkage-temperature bars must be added in the perpendicular direction. The slab does not sit on the beams; rather, concrete around the top of a beam is part of both the beam and the slab. A concrete beam in this situation is considered to be a T-shaped member, with a portion of the slab acting together with the stem of the beam, resulting in a greater structural efficiency and reduced beam depth.

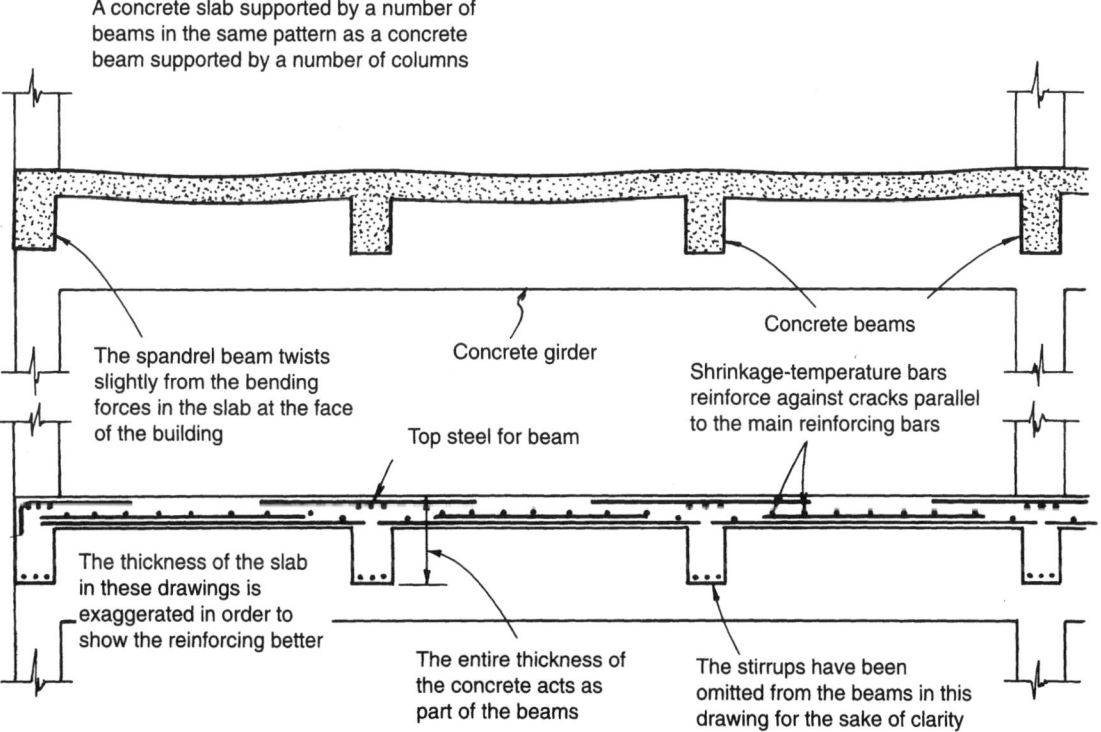

Fig 3.15 Reinforcing for a one-way concrete slab

3.5 A practical approach

In reinforced concrete construction, every floor generally has a beam/slab arrangement and consists of fixed or continuous one-way or two-way slabs that are supported by main and secondary beams. Fig 3.16 on page 51 shows such an arrangement.

The usual arrangement of a slab-and-beam floor is of slabs supported on cross-beams or secondary beams that are parallel to the longer side, with the main reinforcement being parallel to the shorter side. The secondary beams, in turn, are supported on main beams that extend from column to column. Even though beam-to-column connections fall outside the scope of this book, Fig 3.17 on pages 52 to 53 has been inserted to help you understand the concept of merging the beam and the column.

Column/beam junctions are usually highly congested with reinforcement (as Fig 3.17a shows), but if the column shape is the same above and below the junction, detailing and construction problems are fairly straightforward. Column reinforcement should continue through the beams without being cranked, and therefore it is sensible to have columns and beams of the same width. If the column is required to offset or change plan shape between one storey and the next, reinforcement might be impossible to fix, and there might be a risk of poorly compacted concrete in a critical part of the structure. This problem can often be eased by having enough beam depth (as Fig 3.17b shows).

Module 1: Unit 3

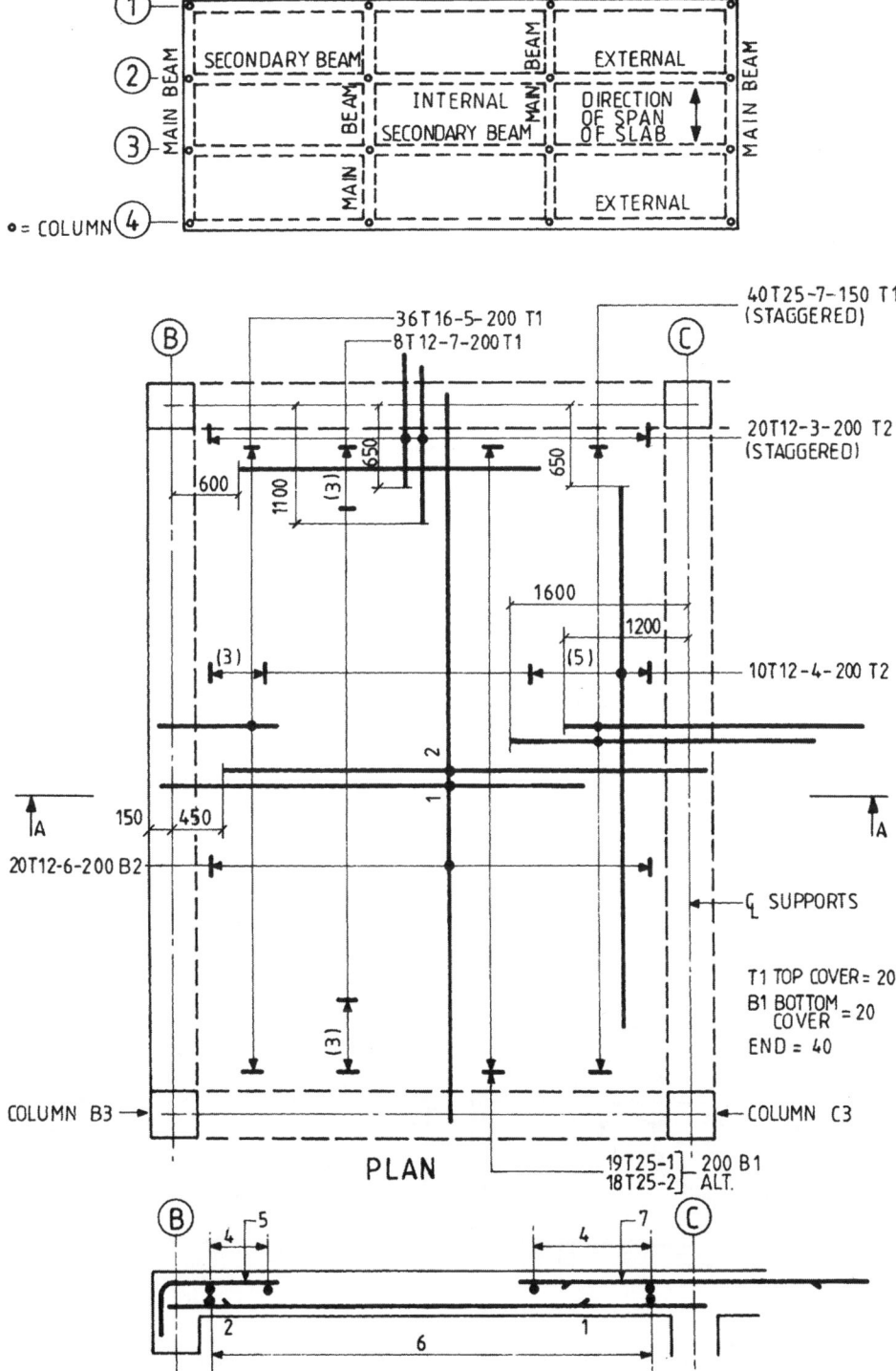

Fig 3.16 Beam and slab arrangement

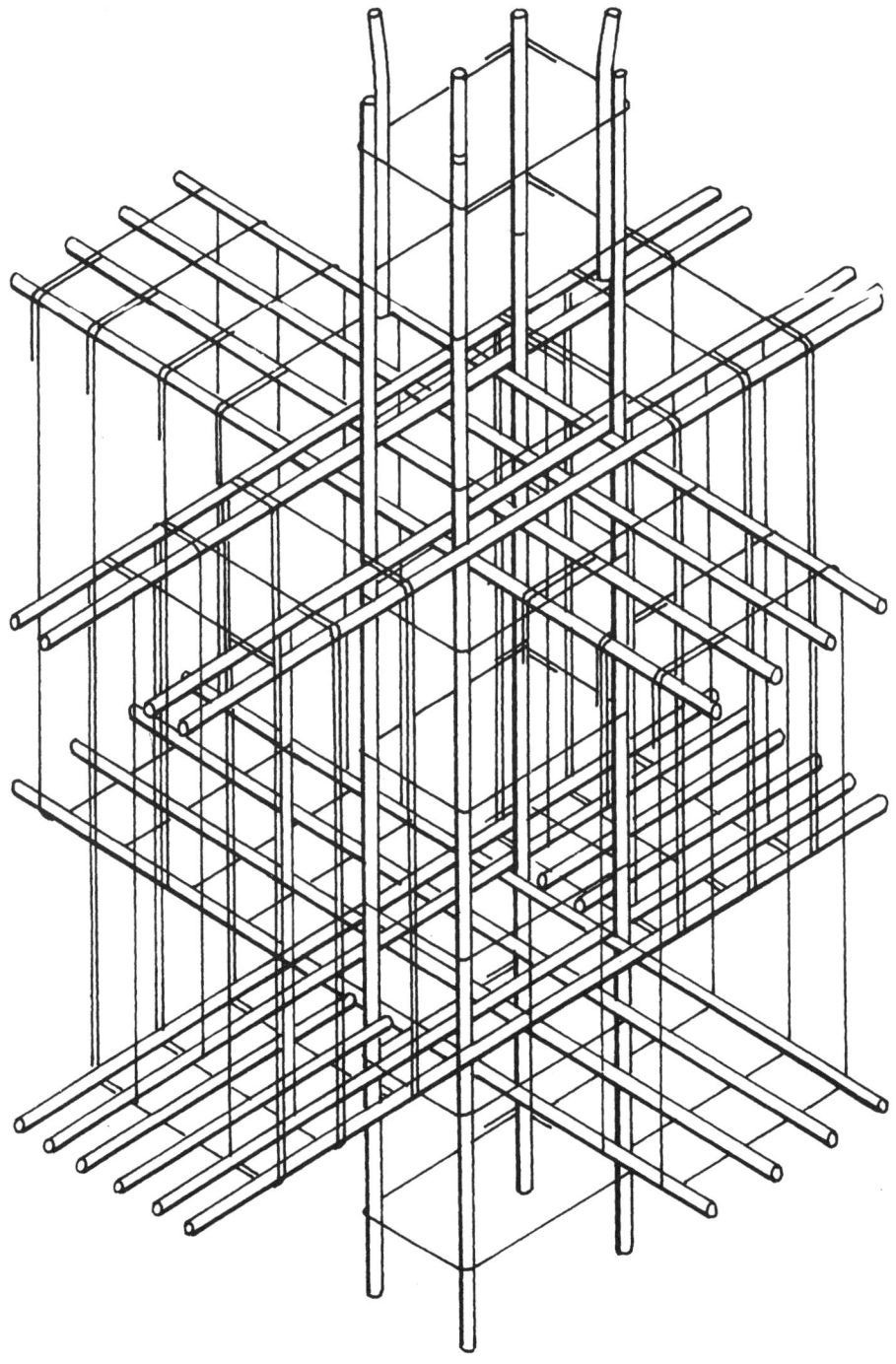

Fig 3.17a A typical reinforcement congestion at beam/column

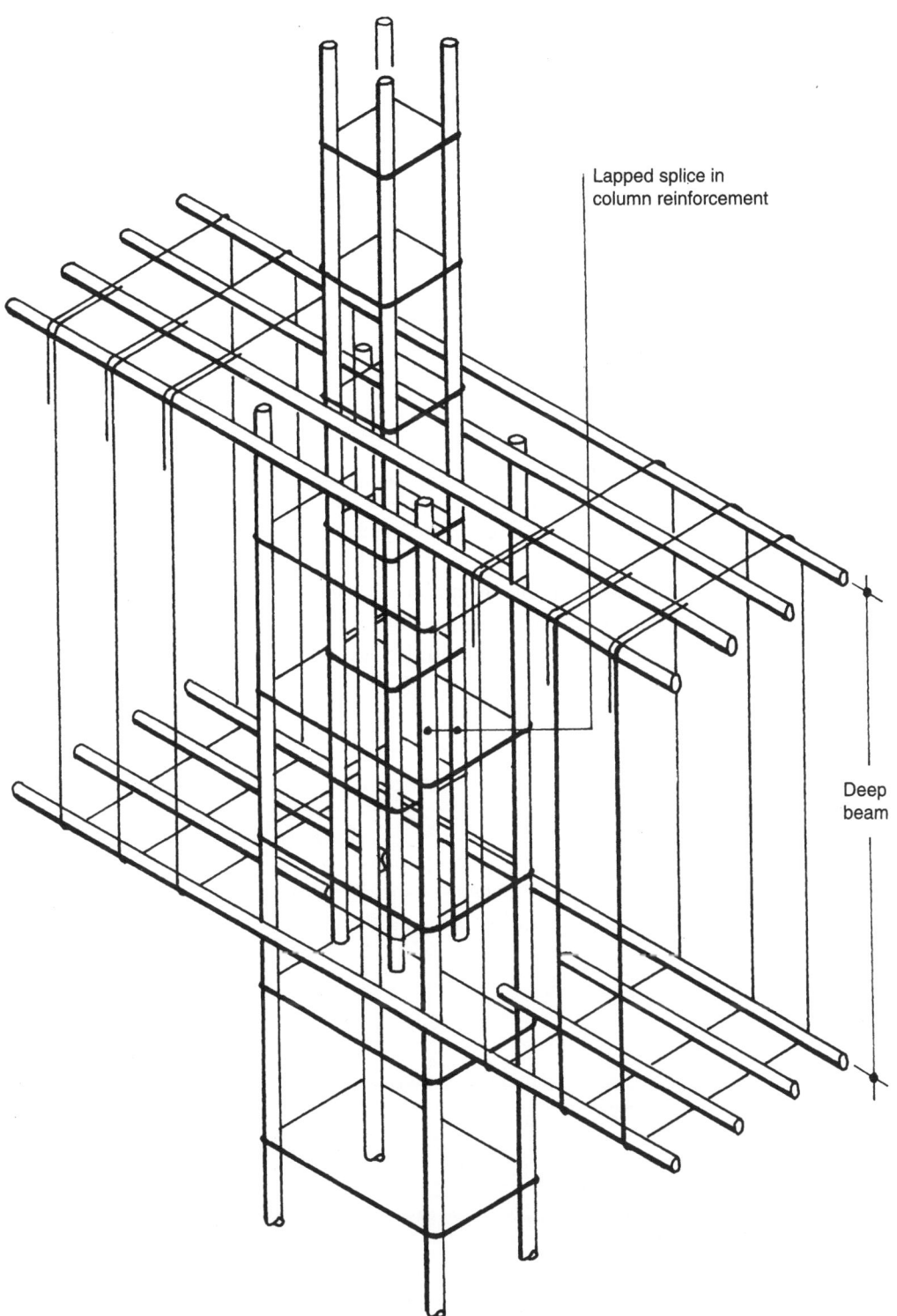

Fig 3.17b A deep beam allows a column splice within its depth

3.6 Summary

Beams are horizontal members of a building, designed to carry external loads, bending moments, shear forces and sometimes torsion moments. Beams may be square, rectangular, flanged or tee, and they can be simply supported, fixed or continuous. Due to the nature of its bending, the simple beam has tensile reinforcement in the bottom and link reinforcement to resist shearing forces across the depth of the beam. This type of beam is usually rectangular in cross-section and its depth usually includes the floor slab.

The floor slab is what you see when walking on the first floor or any floor above ground level. Depending on the type of slab, it is usually supported by beams. There are three general arrangements:
1. One-way spanning slabs: These span between lines of supporting beams and/or walls, which are usually parallel.
2. Two-way spanning slabs: These are supported on a rectangular grid of beams, with columns at beam intersections — the system is not suited to irregular column layouts.
3. Flat slabs: These are carried directly on columns with no beams. This system can accommodate irregular column layouts more easily. Perimeter columns often need to be more closely spaced than internal ones to provide enough stiffness to support the cladding.

SOLVED EXAMPLES

EXAMPLE 3.1 Detailing Beams

Fig 3.18 on page 55 shows a reinforced concrete floor layout for a double-storey building. You are required to detail and schedule:
- the 230×450 simply-supported beam on grid line A; and
- the 230×550 continuous beam on grid line 1.

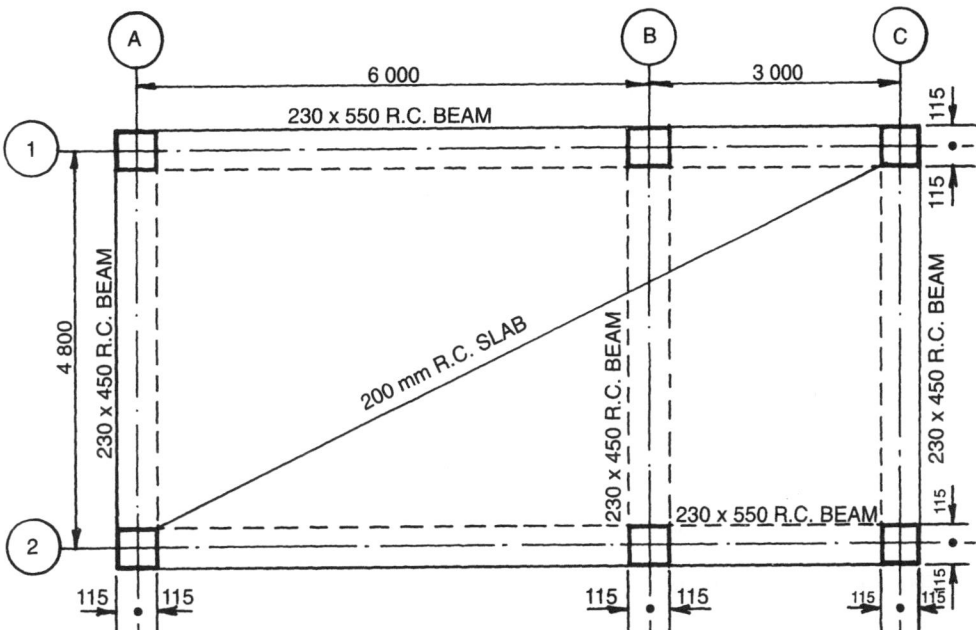

Fig 3.18 Reinforced concrete floor layout: double-storey building

Fig 3.19 on page 56 shows the reinforcement as designed by the engineer. The general notes are as follows:
- Concrete strength 25 MPa;
- Cover: Slab = 20 mm; and
- Beams = 30 mm.

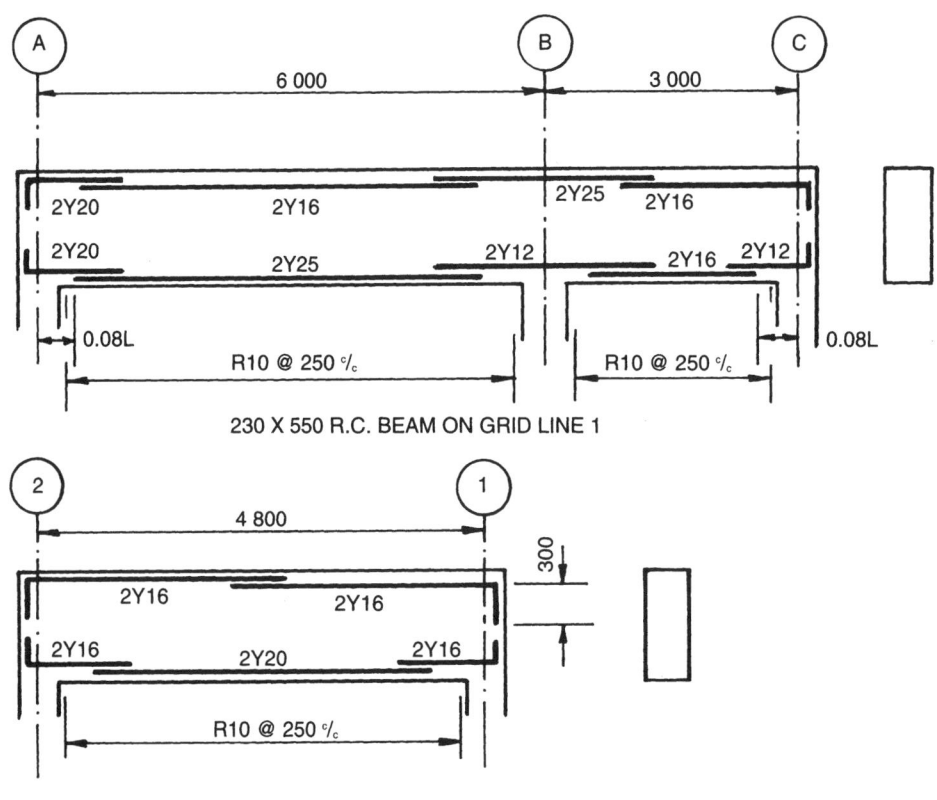

Fig 3.19 Beam reinforcement

Module 1: Unit 3

Solution

(Read in conjunction with Fig 3.20.)

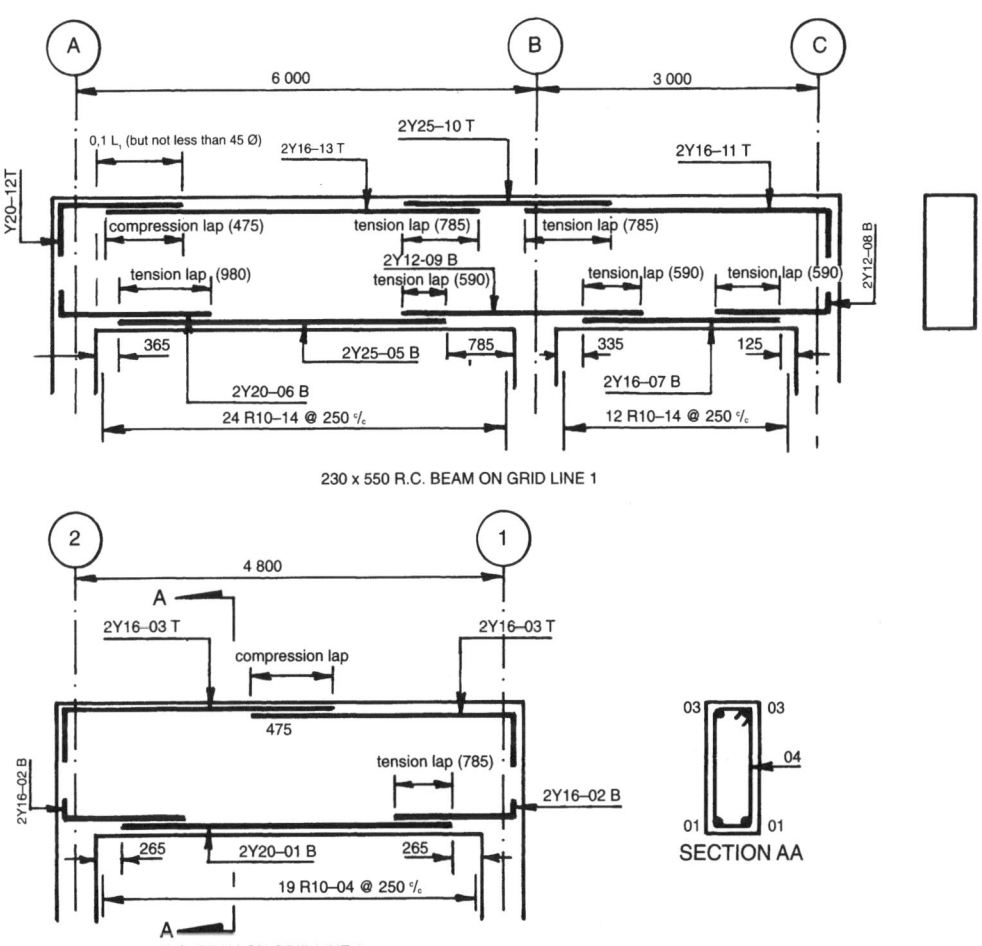

Fig 3.20 Beam reinforcement

Part 1
■ Bar mark 01
To determine the extent of this bar we look at Fig 3.21 on page 58, the bending moment diagram.

57

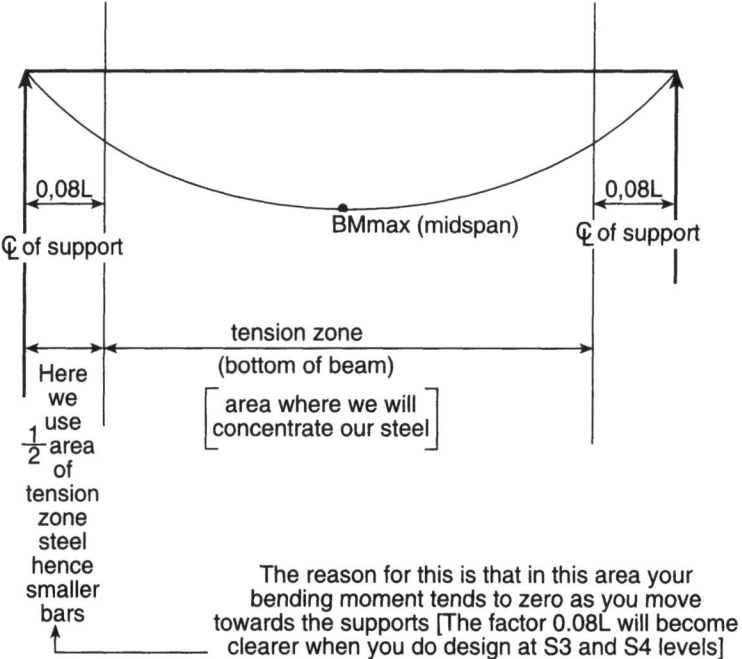

Fig 3.21 Bending moment diagram

Length of bar mark 01
= 4 800 − 2(0,08 × length)
= 4 800 − 2(0,08 × 4 800)
= 4 032
= 4 050
(Round up to the nearest 50.)

Note that you cannot make the bar shorter. This would result in decreasing the length of the bar in the tension zone.

You have to give the steelfixer a dimension to position the bar relative to the face of the column, hence the dimension '265'. The dimension '265' we obtain as follows:

$0.08 L - \frac{1}{2}$ (column width)
= (0,08 × 4 800) − (230)
= 269

It is easier for the steelfixer to measure in multiples of five (5), therefore we round the 269 down to 265. If we rounded up, we would be shortening the bar!

■ Bar mark 02

We use shape code 34.
We calculate the A dimension as follows:
A = tension lap + 265 + column width − beam cover
To determine the tension lap, we use Table 1.1 (Table D.I SABS 0144:1995).

Notice that Table 1.1 on page 3 makes no provision for 25 MPa concrete, so we will interpolate between 20 MPa and 30 MPa.

Tension lap for 20 MPa concrete, bar diameter 16 = 880
Tension lap for 30 MPa concrete, bar diameter 16 = 690
We take the average of 880 and 690 to give us 785.

Therefore, A Length of bar
= 785 + 265 + 230 − 30
= 1250
= A + n
= 1250 + 150
= 1400 (which is a multiple of 50)

■ Bar mark 03

You might wonder why we lap the steel in the compression area since it is not really design steel.

Have you wondered what would happen if the beam were cast slightly longer than its design length? In this case, our top steel (if we had taken it from cover to cover, that is, without a compression lap in the centre of the beam) would result in a reduction in the concrete cover. This is obviously dangerous.

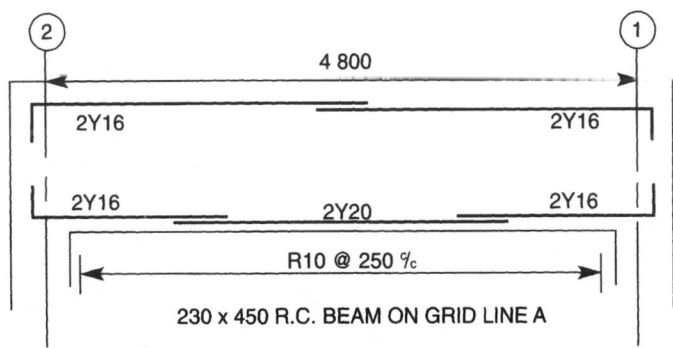

Fig 3.22

Length of bar mark 03 = A + n (shape code 34)
A = y (span) + (column width) − beam cover + (compression lap)
   = (4800) + (230) − 30 + (475)
   = 2722,5

(Remember to use Table 1.1 to interpolate the compression lap for a Ø16 bar using a 25 MPa concrete.)
Round to a multiple of 10, therefore, A = 2730
A + n = 2730 + 150
= 2880

Therefore, we use a length of 2 900. The extra 20 mm will be taken up by the bend, that is, the 'n' dimension. We normally position our stirrups 50 mm from the face of the column. Remember that we try to stay clear of the column steel.

■ **Bar mark 04**
To calculate the number of stirrups required, we need to determine the distance that stretches from the centre line of the first stirrup to the centre line of the last stirrup. We then divide this distance by 250 to give us the number of spaces. Finally, we add one bar to the number of spaces to give us the number of stirrups.

Distance from centre line of first stirrup to centre line of last stirrup
= 4800 − column width − 100 − 10($\frac{1}{2}$ bar diameter)
= 4460

No. of stirrups = $\frac{4460}{250}$ + 1
= 18,84 (Therefore, we use 19 stirrups.)

See if you can calculate the cutting length of the stirrup.

Part 2
Since we are continuing on the same bending schedule, we cannot start detailing the continuous beam by naming the first bar we call up as bar mark 01. However, the mark of the first bar that we call up will continue from the mark of the last bar from Part 1 of this example (that is, the simply supported beam). Therefore, our first bar that we call up on the continuous beam will be bar mark 05.

Before we start detailing and scheduling our continuous beam, let us look at the bending moment diagram in Fig 3.23.

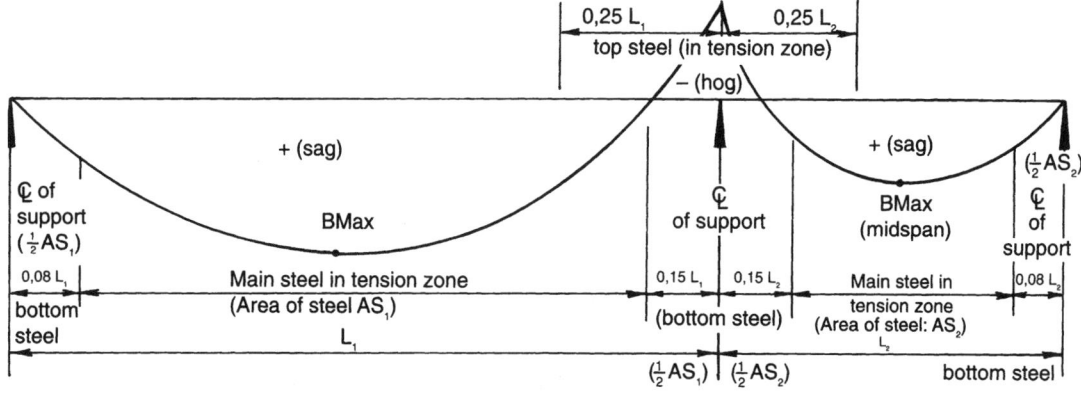

Fig 3.23 Bending moment diagram

The factors 0,08 L, 0,15 L and 0,25 L are used to determine the extensions of the main steel in the tension zones.

0,15 L is used to determine the extent of the bottom steel at the internal supports.
0,25 L is used to determine the extent of the top steel over the supports.
0,08 L is used to determine the extent of the bottom steel at the end supports.

You will see how we arrive at these factors when you do Design at S3 and S4 levels.

Let us see how we use these factors to determine the lengths of our steel:

- **Bar mark 05**

Cutting length = 6 000 − (0,08 × 6 000) − (0,15 × 6 000)
$\qquad$ = 4 620 (Remember to round up to a multiple of 50.)
Therefore, cutting length = 4 650

Note that we must also dimension the extent of the bar by referencing the bar from the face of the columns — hence the dimensions 365 and 785.

- **Bar mark 06**

We use shape code 34.
A = column width − beam cover + 365 + tension lap
$\quad$ = 230 − 30 + 365 + 980

Were you able to interpolate the tension lap from your lap length table?

$\quad$ = 1 545 (We round all bending dimensions to a multiple of 10.)
Therefore, A = 1 550

Cutting length = A + n
$\qquad$ = 1 550 + 170
$\qquad$ = 1 720
Therefore, we use a cutting length of 1 750. (The extra 30 mm will be taken up by the right-angled bend.)

- **Bar mark 07**

Cutting length = 3 000 − (0,15 × 3 000) − (0,08 × 3 000)
$\qquad$ = 2 310
Therefore, cutting length = 2 350

Drawing for Civil Engineering

 Remember to reference the extent of the bar — hence the dimensions 335 and 125.

- **Bar mark 08**

We use shape code 34.
A = tension lap + 125 + 230 − 30
  = 590 + 325
  = 915 (Make A 920.)

Cutting length = A + n
        = 920 + 130
        = 1 050

- **Bar mark 09**

Cutting length = 2 (tension lap) + 785 + 230 + 335
        = 2 530 (Therefore, we use a cutting length of 2 550.)

- **Bar mark 10**

Cutting length = (0,25 × 6 000) + (0,25 × 3 000)
        = 2 250

- **Bar mark 11**

We use shape code 34.
A = 3 000 + $\frac{1}{2}$ (col width) − beam cover − (0,25 × 3 000) + tension lap
  = 3 120

Cutting length = A + n
        = 3 120 + 150
        = 3 270

Therefore, we use 3 300 as our cutting length.

- **Bar mark 12**

Note that the designer decides whether this bar is required or not. If the bar is required, the A dimension is governed by the design criteria as given on the drawing.

 This will become clear in Design at S3 and S4 levels.

Therefore, before we can calculate the A dimension, we must first calculate:
0,1 L = 0,1 × 6 000
    = 600
and 450 = 45 × 20
     = 900
We therefore use the greater value, that is, 900.

Table 3.2 Bending schedule

Member Mark; Size; No. Off.	Reinforcement				Shape code	Bending dimensions				EorR	Mass kg	Fixing and non-standard bending details
	No. ea	Total no.	Size	Mark	Length (total length in mm)		A	B	C	D			
1 230 × 450 × 5030 (3 No. Off)	2	6	Y20	01	4050	20	4050						
		12	Y16	02	1400	34	1250	150					
		12	Y16	03	2900	34	2730	170					
		57	R10	04	1300	60	390	170	180				
2 R.C. Beam 230 × 550 × 9230 (2 No. Off)	2	4	Y25	05	4650	20	4650						
		4	Y20	06	1750	34	1550	200					
		4	Y16	07	2350	20	2350						
		4	Y12	08	1050	34	920	130					
		4	Y12	09	2550	20	2550						
		4	Y25	10	2250	20	2250						
		4	Y16	11	3300	34	3120	180					
		4	Y20	12	1300	34	1100	200					
		4	Y16	13	4750	20	4750						
		72	R10	14	1530	60	490	170	180				

Therefore, A = 900 + column width − beam cover
= 1 100

You should be able to calculate the cutting length on your own.

- **Bar mark 13**

Are you ready for a challenge? See if you can calculate the length of the bar! Complete the example by doing the following:
1. Draw a section taken through the beam. Remember to indicate on the elevation where your section is taken. Insert all bar marks on your section.
2. Complete the bending schedule by completing the column for mass.

EXAMPLE 3.2 Detailing Floor Slabs

A slab measuring 6 m × 3 m is simply supported on two sides by 230 mm brick walls (that is, it is a one-way spanning slab). The slab is 200 mm thick and is reinforced at the bottom as follows:
- Short span (main steel) — Y12 (shape code 34) at 125 mm centres (that is, parallel to short span);
- Long span (distribution steel) — Y10 (shape code 20) at 250 mm centres (that is, parallel to long span);
- Concrete cover is 30 mm; and
- Concrete strength is 40 MPa.

Use a scale of 1:20 to detail (in plan and section) the reinforcement layout and prepare the bending schedule.

Solution

(Read in conjunction with Fig 3.24 on page 65.)

Module 1: Unit 3

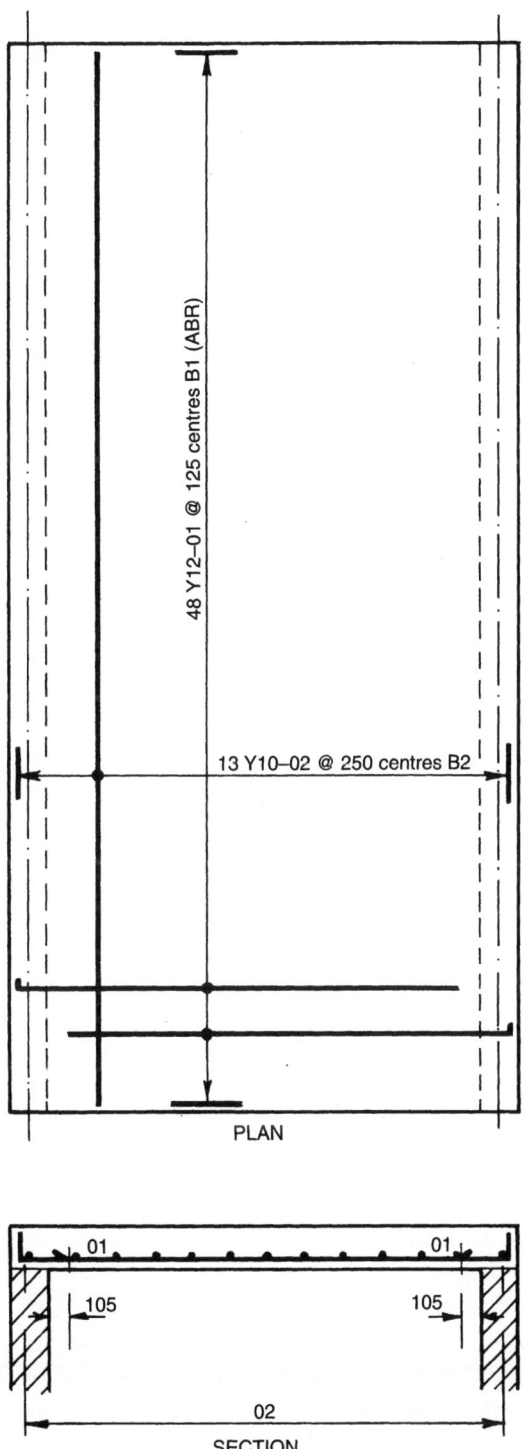

Fig 3.24 Simply-supported slab

Like simply-supported beams, we base the reinforcement design on the maximum bending moment, which occurs at mid-span. Therefore, the main area of steel is concentrated in the area bordered by 0,08 L, from the centre of the left support to 0,08 L from the centre of the right support. Because the bending moment tends towards zero as we move towards the supports, it is economical to reduce the amount of reinforcement to 50% of the main area of steel required, hence the idea to reverse each bar.

Can you see that we have main steel (that is, the straight part of the shape code 34 bar) running every 125 mm centre to centre, and the L-shaped piece running every 250 centres (that is, double the main steel centres)? This is what we mean when we talk about reducing the amount of reinforcement to 50% of the main steel. We use straight bars as distribution steel, because we have no supports along the short span. Therefore, it is not necessary to bend the ends up into the slab. You should be in a position to calculate the number of bars required in both directions, since we have been through these calculations in previous examples.

Pay particular attention to the way we call up the 01 bar mark. We indicate that we reverse every alternate bar by using the standard abbreviation (ABR). In addition, we give the steelfixer the dimension from the face of the column to the end of the 01 bar mark. The resulting dimension should be rounded down to a dimension divisible by 5 mm; if you round up, you shorten the bar.

Let us look at the calculation of the bar lengths.

■ **Bar mark 01**
A = 3 000 − 230 − 105 − 30
　= 2 635 (We use 2 640.)

The dimension for the span = 3 000 − 2(y column width) n = 130
Note: We must check whether we can accommodate this dimension without reducing the cover. We do this by deducting the cover at the top and bottom of the slab from the slab thickness, that is, 200 − 30 − 30 = 140. This gives a dimension greater than n = 130.

Therefore, we can use n = 130.
Cutting length = A + n
　　　　　　 = 2 640 + 130
　　　　　　 = 2 770
(If we round this up to 2 800, we increase the n dimension by 30 mm, which is not permissible. Therefore, we rather increase the A dimension by 30 mm. So A becomes 2 670.)

Always remember to carry out this check or else you might land up in trouble.

Bar mark 02

Cutting length = 6 000 − 30 − 30
             = 5 940

(If we round this length up to 5 950, we will be once again reducing the cover. The solution is to round down to 5 900.) Again, it is important to carry out this check instead of just performing the calculations mechanically. Think about what you are doing and why. Your bending schedule should look as follows:

Table 3.3 Bending schedule

Member	Reinforcement					Bending dimensions					Mass
Mark; Size; No. Off	No. ea.	Total no.	Size	Mark	Length (total length in mm)	Shape code	A	B	C	D	kg
Slab 6 000 × 3 000 × 200 (1 off)	48	48	Y12	01	2 800 (134 400)	34	2 670				
	13	13	Y10	02	5 900 (76 700)	20					

You can finish off the bending schedule by completing the 'mass' column.

EXAMPLE 3.3

A slab measuring 6 m × 3 m is simply supported on four sides by 230 mm brick walls (that is, it is a two-way spanning slab). The slab is 200 mm thick and is reinforced at the bottom as follows:

- Short span — Y12 (shape code 34) at 125 mm centres (that is, parallel to short span);
- Long span — Y10 (shape code 34) at 250 mm centres (that is, parallel to long span);
- Concrete cover is 30 mm; and
- Concrete strength is 40 MPa.

Use a scale of 1:20 to detail (in plan and section) the reinforcement layout and prepare the bending schedule.

Solution

(Read in conjunction with Fig 3.25.)

345
48 Y12–01 @ 125 centres B1 (ABR)
13 Y10–02 @ 250 centres B2 (ABR)
345
PLAN
01
01
105
105
02
SECTION

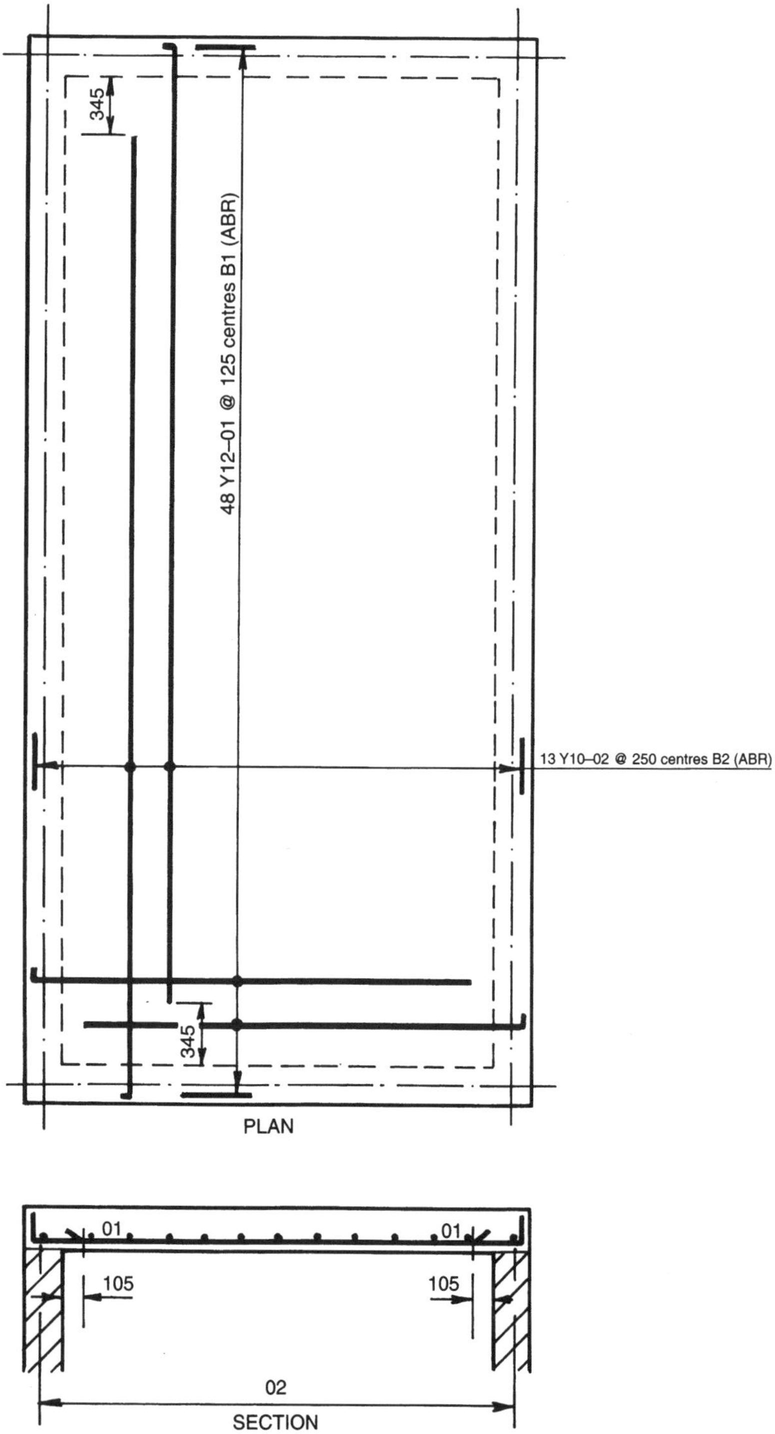

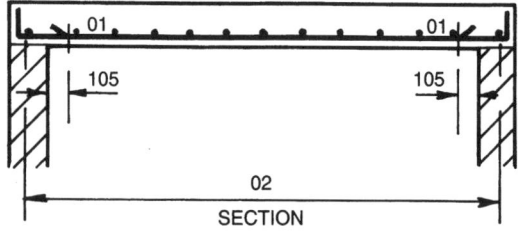

Fig 3.25

You might wonder why the steel in the long span has changed from straight bars (as in the case of the one-way spanning slab) to shape code 34 bars that are cut off (or curtailed) in the same way as the short span bars. Well, the reason should be obvious. We have supports in both directions; therefore, we have bending moments in both directions. We therefore base the reinforcement design on the maximum bending moment in both directions and reduce the amount of reinforcement to 50% as the bending moment tends towards zero moving towards the supports.

■ **Bar mark 01**
All information remains the same as in Example 3.2.

 Do you agree?

■ **Bar mark 02**
1. Offset from column face $= 0{,}08\,L - \frac{1}{2}$ column width
$$= 0{,}08(5\,770) - 115$$
$$= 346{,}6 \text{ (Use 345.)}$$

2. $A = 6\,000 - 230 - 345 - 30$
   $= 5\,395$ (Use 5 400.)

3. $\quad\quad\quad\quad n = 130$
Cutting length $= A + n$
$\quad\quad\quad\quad\quad = 5\,400 + 130$
$\quad\quad\quad\quad\quad = 5\,530$
(If we round up to 5 550, we increase n by 20 mm. Therefore, we adjust A to 5 420.)

Bending schedule is given by:

Table 3.4 Bending schedule

Member	Reinforcement					Bending dimensions					Mass
Mark; Size; No. Off	No. ea	Total no.	Size	Mark	Length (total length in mm)	Shape code	A	B	C	D	kg
Slab 6 000 x 3 000 x 200	48	48	Y12	01	2 800 (134 400)	34	2 670				
(1 off)	13	13	Y10	02	5 550 (72 150)	34	5 420				

Drawing for Civil Engineering

You can finish off the bending schedule by completing the 'mass' column.

EXAMPLE 3.4

Fig 3.26 shows a slab simply supported on two sides by 230 mm × 500 mm deep reinforced concrete beams. This means that it is a one-way spanning slab. The slab and beams are monolithic (cast as a unit). The beams are in turn supported at their ends by four 230 mm × 230 mm reinforced columns. The slab is 170 mm thick and is reinforced at the bottom as follows:

- Short span (main steel) — Y12 (shape code 20) at 220 mm centres (that is, parallel to short span);
- Short span (lapping on to the main steel) — Y12 (shape code 38) at 440 mm centres;
- Provide Y10 lacer bars (also referred to as lacing bars);
- Long span (distribution steel) — Y10 (shape code 20) at 280 mm centres (that is, parallel to long span);
- Concrete cover is 25 mm (slab) and 30 mm (beam); and
- Concrete strength is 30 MPa (slab).

Use a scale of 1:20 to detail (in plan and section) the reinforcement layout and prepare the bending schedule.

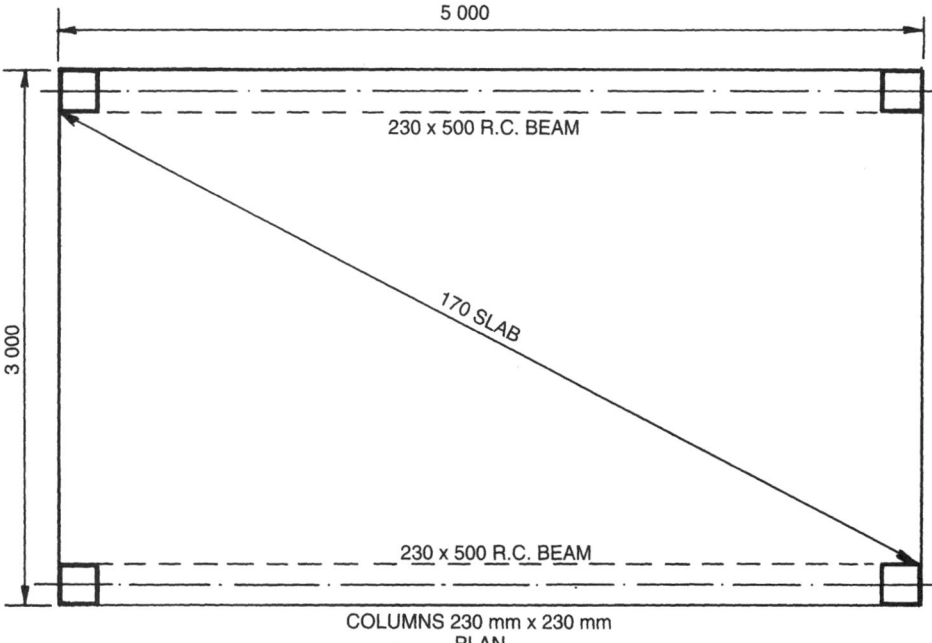

Fig 3.26 A slab simply supported on two sides

Solution
(Read in conjunction with the version of Fig 3.27 which has been drawn to

scale and can be found in the pocket at the back of this book. The version of Fig 3.27 which can be found on page 72 is not drawn to scale and is intended for quick reference only.)

Let us discuss the difference between this slab (supported by beams) and the slab in Example 3.1 (supported by walls). We design the slab supported by walls to counteract shear and bending. However, the slab supported by beams is designed to counteract shear and bending as well as torsion. Since it is cast as part of the beams, there is a tendency for the slab to twist at the intersection. To counteract the torsion, we can provide reinforcement as shown at the intersection of the beam and slab of Fig 3.16 on page 51. However, to make the reinforcement design easier, we can also lap a bar (shape code 38) directly onto every second main steel. This bar (that is, shape code 38) will serve a dual purpose. It will counteract the tendency for the twist at the beam slab intersection, and simultaneously reduce the amount of reinforcement required to 50%, as the bending moment tends to zero moving towards the supports.

You will notice that in the plan view, we have drawn the bar mark 02 as we would see it in section. It is true that in the plan view we actually see this bar as a straight line. To assist the steelfixer when reading the plan, we draw the bar as shown, to distinguish it from a straight bar.

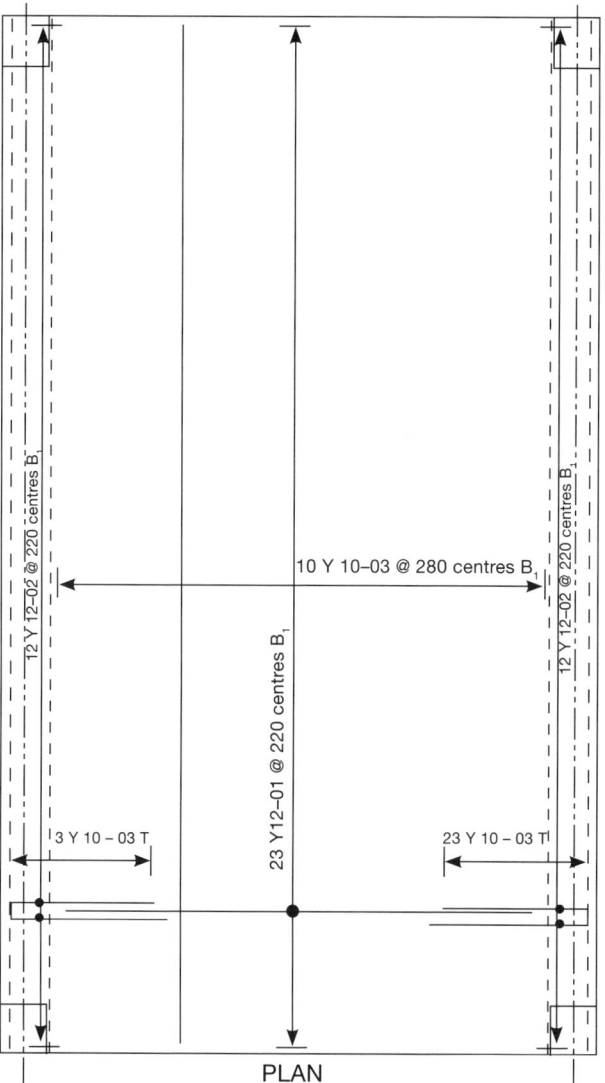
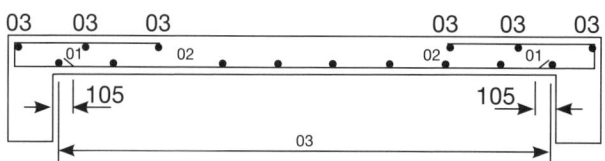

Fig 3.27 (See scale drawing in the pocket at back of book.)

Module 1: Unit 3

 A clever solution, don't you think?

Let us go through the calculations together.

- **Bar mark 01**

We fix these bars from the front cover of the slab to the back cover of the beam. Therefore, the number required

$= [\dfrac{5\,000 - 2(25) - 2(½ \text{ bar diameter})}{220}]$

$= 23{,}44$ (We will use 23 bars.)

Bar length $= 3\,000 - 2(230) - 2(\text{offset from the inside face of the beam})$
$\qquad\qquad = 3\,000 - 460 - 2[0{,}08(2\,770) - \dfrac{1}{2} \text{ beam width}]$
$\qquad\qquad = 2\,540 - 2(105)$ (Note: 105 rounded down from 106,6)
$\qquad\qquad = 2\,330$ (We use 2 350.)

- **Bar mark 02**

These bars lap onto every second '01' bar. Therefore, we use 12 bars (half the number of '01' bars. If you refer to Table 5 of SANS 82, you will notice that there are two sketches for shape code 38. We will use the second sketch with the A dimension lapping onto our main bar, the B dimension bending up into the beam and the C dimension positioned at the top of the beam/slab intersection. You will be able to see this very clearly in the section.

$A = \text{tension lap} + 105 + 230 - \text{beam cover}$
$\quad = 520 + 105 + 230 - 30$
$\quad = 825$ (We use 830.)

$B = \text{slab thickness} - 2(\text{slab cover})$
$\quad = 170 - 50$
$\quad = 120$

To calculate the C dimension, we use the same factors we used in the example of the continuous beam (bar mark 12).

First, calculate $0{,}1\,L = 0{,}1(3\,000 - 230)$
$\qquad\qquad\qquad\qquad = 277$
Next, we calculate $45 \times \text{diameter of bar} = 45 \times 12$
$\qquad\qquad\qquad\qquad\qquad\qquad\qquad = 540$
We use the greater value, which is 540.

Therefore $C = 230 + 540 - \text{beam cover}$
$\qquad\qquad = 740$

The cutting length of bar mark 02 = A + B + C − r − 2d
= 830 + 120 + 740 − 36 − 2(12)
= 1 630 (We round up to 50.)
= 1 650

■ Bar mark 03

These are your distribution bars. You will notice that the first bar starts 50 mm away from the inside face of the left beam. The last bar ends 50 mm away from the inside face of the right beam. The reason for this is twofold. First, the slab width starts at the inside face of the left beam and ends at the inside face of the right beam. So the slab dimension is 2 540 × 5 000. In addition, we already have reinforcement in the beams.

Number of bars = 10 cutting length
= 5 000 − 2(slab cover)
= 4 950

Lacing bars

We usually use three bars on either end of the slab/beam intersection. Can you see that the length of these bars will be the same as the '03' bar? We can use the same bar mark because the shape code is also the same. We must not forget to add the total number of '03' bars to the total number of lacing bars when we complete the bending schedule. You might wonder why we draw the lacing bars as dashed lines in the plan view. This is to indicate that these bars occur in the top of the slab.

Table 3.5 Bending schedule

Member	Reinforcement					Bending dimensions					Mass
Mark; Size; No. Off	No. ea.	Total no.	Size	Mark	Length (total length in mm)	Shape code	A	B	C	D	kg
Slab 5 000 × 2 540 × 170 (1 off)	23	23	Y12	01	2 350 (54 050)	20					
	24	24	Y12	02	1 600 (39 600)	38	830	120			
	16	16	Y10	03	4 950 (79 200)	20					

You can complete the calculations for the mass on your own.

On your bending schedule, you only schedule the bending dimensions shown in Table A1 of SANS 282:2004. For this reason we do not schedule the C dimension of the '02' bar.

Module 1: Unit 3

EXAMPLE 3.5

Fig 3.28 shows a slab simply supported on four sides by 230 mm × 500 mm deep reinforced concrete beams (that is, it is a two-way spanning slab). The slab and beams are monolithic (cast as a unit). The beams are in turn supported at their ends by four 230 mm × 230 mm reinforced columns. The slab is 170 mm thick and is reinforced at the bottom as follows:
- Short span (main steel) — Y12 (shape code 20) at 220 mm centres (that is, parallel to short span);
- Short span (lapping on to the main steel) — Y12 (shape code 38) at 440 mm centres;
- Provide Y10 lacer bars (also referred to as lacing bars);
- Long span — Y10 (shape code 20) at 280 mm centres (that is, parallel to long span);
- Y10 (shape code 38) lapping on to every second Y10 (shape code 20) bar;
- Provide Y10 lacer bars;
- Concrete cover is 25 mm (slab) and 30 mm (beam); and
- Concrete strength is 30 MPa (slab).

Use a scale of 1:20 to detail (in plan and section) the reinforcement layout and prepare the bending schedule.

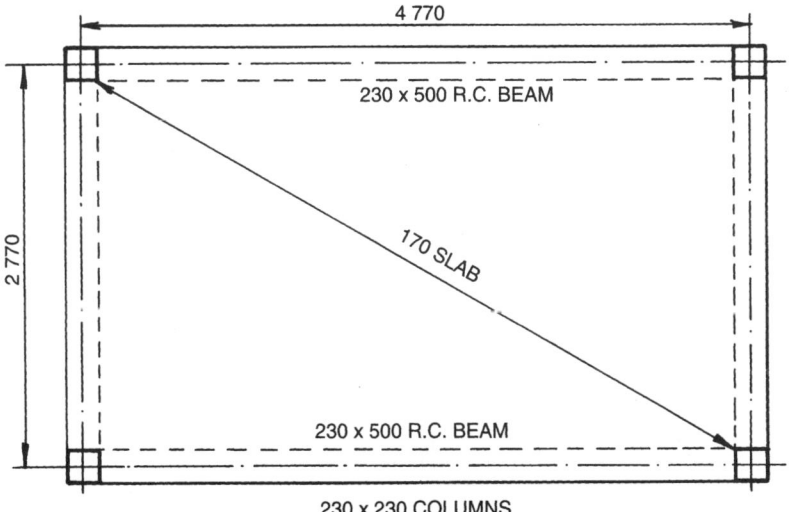

Fig 3.28 A slab simply supported on four sides

Solution
In this example you have the slab supported in both directions; thus, bending, shear and torsion occur in both directions. Therefore, the reinforcement design (namely steel designed as in the short span of Example 3.4) will apply in both directions. Before you look at the solution (given in Fig 3.29 and the bending schedule in

Drawing for Civil Engineering

Table 3.6), first try to work through the example on your own. The version of Fig 3.29 which can be found on this page is not drawn to scale, and is for quick reference only. For the solution to Example 3.5, consult the version of Fig 3.29 which has been drawn to scale and can be found in the pocket at the back of the book.

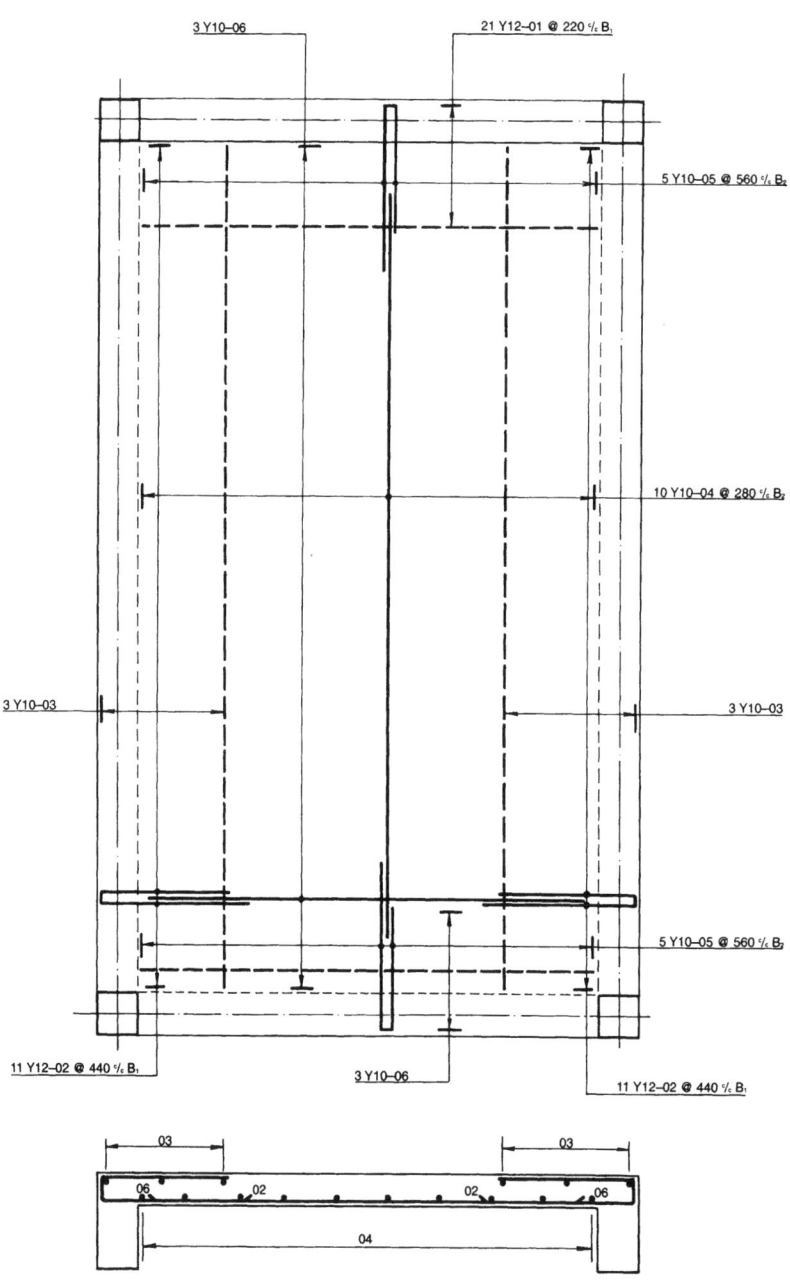

Fig 3.29 (See scale drawing in the pocket at back of book.)

The dimensions of your slab are now different to those of Example 3.4.

Table 3.6 Bending schedule

Member	Reinforcement					Bending dimensions				Mass
Mark; Size; No. Off	No. ea.	Total no.	Size	Mark	Length (total length in mm)	Shape code	A	B	C	D	kg
Slab 5 000 × 2 540 × 170 (1 off)	21	23	Y12	01	2 350 (54 050)	20				
	22	24	Y12	02	1 650 (36 300)	38	830	120		
	6	16	Y10	03	4 050 (40 500)	20				
	10	10	Y10	04	1 650 (16 500)	38	900	120		
	6	6	Y10	06	2 450 (14 700)	20				

Note: Cutting length of bar mark 03 = 4 770 − 2(115) − 2(25)
= 4 490
(We round down to 4 450. To round up to 4 500 would reduce the slab cover.)
C = 230 − beam cover + (the greater of 0,1 L and 45 × bar diameter)
= 200 + (the greater of 477 and 450)
= 677 (We use 680.)

Cutting length of bar mark 04 = 2 770 − 2(115) − 2(25)
= 2 490

(We round down to 2 450 because the slab cover would be reduced if we rounded up to 2 500.)

We do not provide the solutions to the following activities in this book. Ask your lecturer to check your attempts.

Activity 3.1

A slab measuring 5 m × 3,5 m is simply supported on two sides by 230 mm brick walls (that is, it is a one-way spanning slab).

The slab is 170 mm thick and is reinforced at the bottom as follows:
- Short span (main steel) — Y12 (shape code 34) at 200 mm centres (that is, parallel to short span);
- Long span (distribution steel) — Y10 (shape code 20) at 300 mm centres (that is, parallel to long span);
- Concrete cover is 20 mm; and
- Concrete strength is 20 MPa.

Use a scale of 1:20 to detail (in plan and section) the reinforcement layout and prepare the bending schedule.

Activity 3.2

A slab measuring 6,5 m × 3,5 m is simply supported on four sides by 230 mm brick walls (that is, it is a two-way spanning slab). The slab is 180 mm thick and is reinforced at the bottom as follows:
- Short span — Y12 (shape code 34) at 175 mm centres (that is, parallel to short span);
- Long span — Y10 (shape code 34) at 250 mm centres (that is, parallel to long span);
- Concrete cover is 25 mm; and
- Concrete strength is 25 MPa.

Use a scale of 1:20 to detail (in plan and section) the reinforcement layout and prepare the bending schedule.

Activity 3.3

Fig 3.30a on page 79 shows a slab simply supported on two sides by 300 mm × 500 mm deep reinforced concrete beams (that is, it is a one-way spanning slab). The slab and beams are monolithic (cast as a unit). The beams are in turn supported at their ends by four 300 mm × 300 mm reinforced columns. The slab is 180 mm thick and is reinforced at the bottom as follows:
- Short span (main steel) — Y12 (shape code 20) at 250 mm centres (that is, parallel to short span);
- Short span (lapping on to the main steel) — Y12 (shape code 38) at 500 mm centres;
- Provide Y10 lacer bars (also referred to as lacing bars);
- Long span (distribution steel) — Y10 (shape code 20) at 220 mm centres (that is, parallel to long span);
- Concrete cover is 25 mm (slab) and 35 mm (beam); and
- Concrete strength is 30 MPa (slab).

Use a scale of 1:20 to detail (in plan and section) the reinforcement layout and prepare the bending schedule.

Module 1: Unit 3

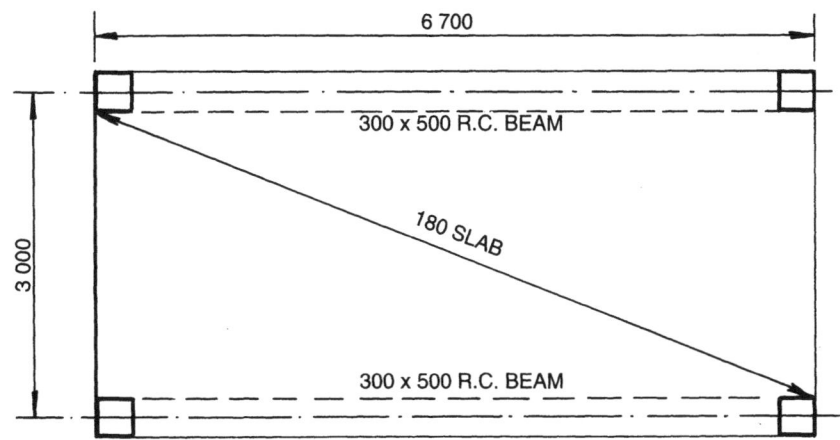

Fig 3.30a Simply-supported slab on two sides

Activity 3.4

Fig 3.30b shows a slab simply supported on four sides by 300 mm × 500 mm deep reinforced concrete beams (that is, it is a two-way spanning slab). The slab and beams are monolithic (cast as a unit). The beams are in turn supported at their ends by four 300 mm × 300 mm reinforced columns.

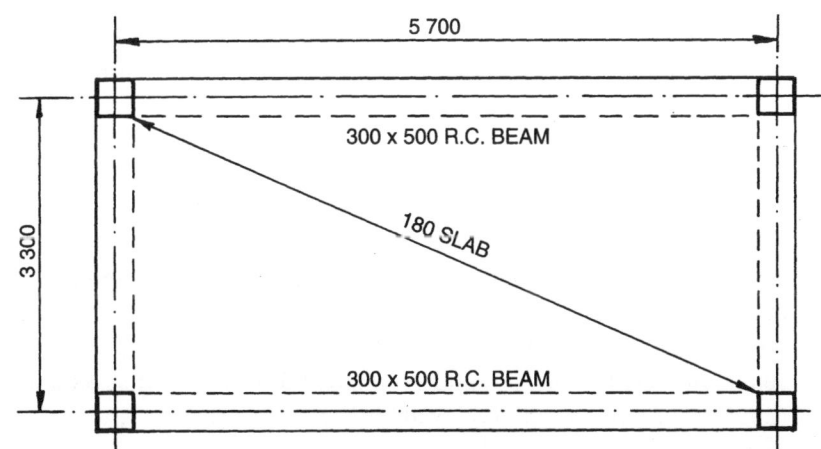

Fig 3.30b Simply-supported slab on four sides

The slab is 180 mm thick and is reinforced at the bottom as follows:
- Short span (main steel) — Y12 (shape code 20) at 250 mm centres (that is, parallel to short span);
- Short span (lapping on to the main steel) — Y12 (shape code 38) at 500 mm centres;

Drawing for Civil Engineering

- Provide Y10 lacer bars (also referred to as lacing bars);
- Long span ± Y10 (shape code 20) at 250 mm centres (that is, parallel to long span);
- Y10 (shape code 38) lapping on to every second Y10 (shape code 20) bar;
- Provide Y10 lacer bars;
- Concrete cover is 25 mm (slab) and 30 mm (beam); and
- Concrete strength is 30 MPa (slab).

Use a scale of 1:20 to detail (in plan and section) the reinforcement layout and prepare the bending schedule.

Activity 3.5

Figure 3.31a on page 81 shows a plan view of the first floor of a double-storey building. The steel (as designed by the engineer) is also shown on the plan view. Figure 3.31b on page 81 shows the reinforcement design for the beams.

1. Draw the reinforcement layouts of each beam, including a section taken through each of the beams (indicate on your drawing where you have taken your section) and complete the bending schedule.
   - Use a scale of 1:20.
   - Supply all the necessary dimensions on your reinforcement drawings so that the steelfixer can place the reinforcement in the correct position.
2. Draw the reinforcement layout of the continuous slab and complete the bending schedule.

You have been given the curtailment (cut off) for the bars at the internal supports (both for the bottom and for the top bars). Remember that the curtailment at all the external supports remains 0,08 L. Be careful when you calculate the curtailment at the supports; you have two different spans, namely $L_1 = 4\,500$, and $L_2 = 3\,900$. What you must remember is that the bending moment diagram is the same as for continuous beams. Regard all laps as tension laps.

Hints
Bar mark 08 (cutting length = $0{,}3\,L_1 + 0{,}15\,L_2$) alternates with bar mark 09 (cutting length = $0{,}15\,L_1 + 0{,}3\,L_2$)

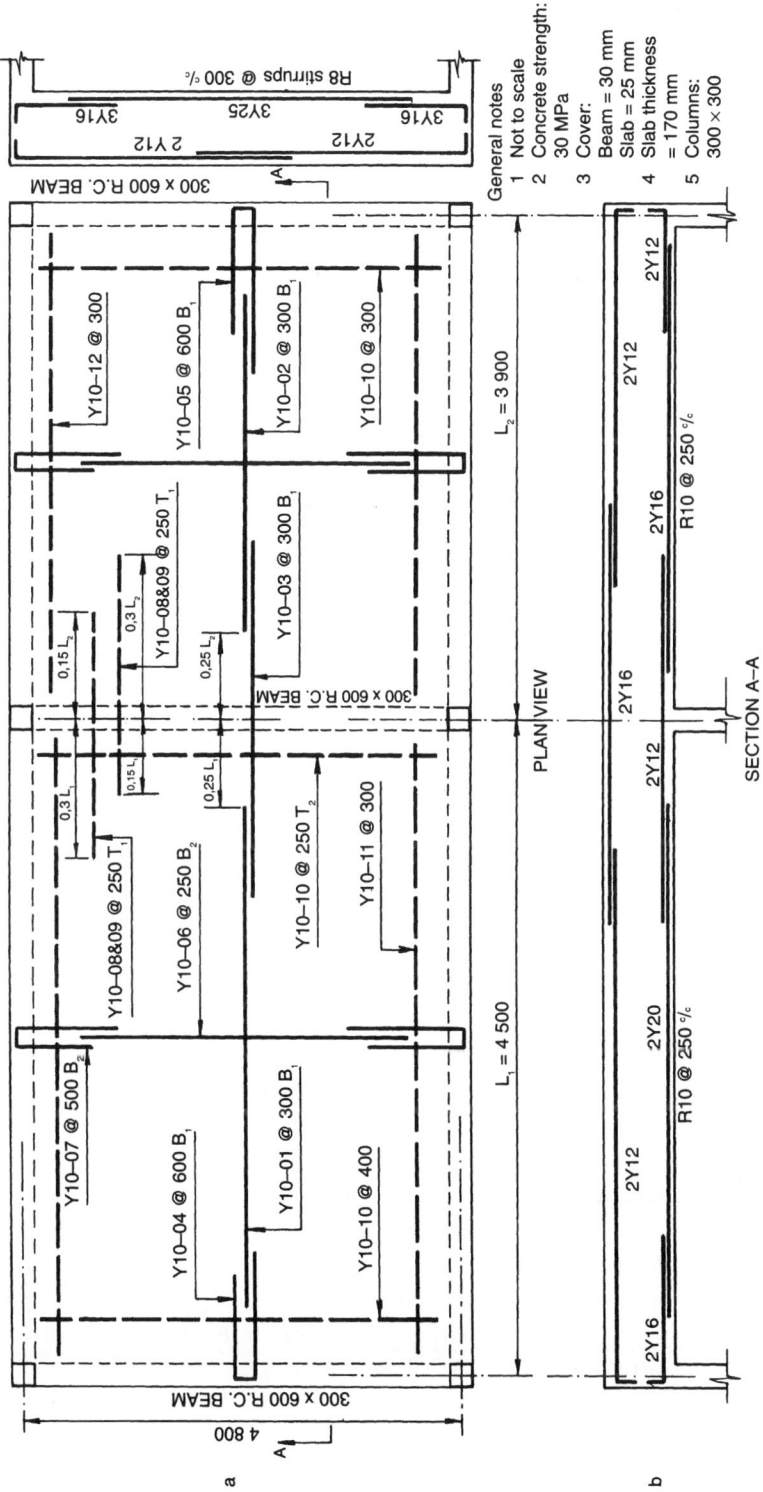

Fig 3.31 Plan view of steel detailing

Bar mark 06 is not fixed inside the internal supports. (Remember that you have beam reinforcement!)

■ **Bar mark 10**

The first bar starts $0,3\ L_1$ to the left of the internal support, with the last bar ending $0,3\ L_2$ to the right of the internal support.

Regard this activity as a project. You might have to ask your lecturer to assist you, but first compare the activities on slabs with the activities on beams. Good luck!

Self-evaluation

1. Describe a reinforced concrete beam.
2. Which shapes are generally used for beams?
3. Where will the maximum bending moment occur in beams?
4. What is a point of contra flexure in beams?
5. What is the meaning of 'hanger bars' in beams?
6. When will you use bent-up bars in beams?
7. What is the difference between suspended slabs and supported slabs?

SELF-EVALUATION ANSWERS
1. Reinforced concrete beams are structural elements designed to carry external loads, such as slabs.
2. The shapes generally used for beams are square, rectangular, flanged or T-shaped.
3. The maximum bending moment occurs at the centre of a beam.
4. When drawing the bending moment diagram, a situation might occur where a positive bending moment changes to a negative bending moment. The point where the bending moment is zero is where the point of contra flexure occurs.
5. Hanger bars are bars onto which stirrups have to be hung to form a rigid cage in a beam to prevent collapse when concrete is placed.
6. You would use bent-up bars in beams when the stirrups alone cannot counteract the shearing forces.
7. Suspended slabs are supported on edges of beams and walls that are supported on columns (known as flat slabs). Supported slabs may be one-way slabs (main reinforcement in one direction) only or two-way slabs (reinforcement in two directions).

Module 1: Unit 4

Unit 4 Using Computer Aided Concrete Training

For this unit you will need the assistance of a lecturer or somebody acquainted with CAD software.

4.1 Introduction to COMPACT

COMPACT is a suite of computer-aided learning programs that cover 11 topics on concrete technology and the design of concrete structures. Produced primarily for undergraduate use, the interactive programs use high-quality photographs and graphics to present teaching material in an easily absorbed and interesting manner. Several of the modules, particularly on reinforced concrete design, will also appeal to graduate engineers.

As this is a British program, the design was done according to British Standards and the student should apply his/her knowledge in South Africa according to South African Standards when dealing with the activities in Unit 4.

Topics covered

(Excerpt from http://compact.shef.ac.uk) The following page is a list of the topics covered in each of the individual COMPACT modules. The topic that we will mostly deal with in this book is 4.1.8, Drawing and detailing of Concrete Structures.

4.1.1 Advanced Design of Reinforced Concrete Structures

- Introduction;
- Behaviour;
- Limit States;
- Material Properties;
- Loads;
- Design Process;
- Analysis of Structures;
- Moment Redistribution;
- Analysis of Sections;
- Flexure;
- Combined Axial Force and Flexure; and
- Shear.

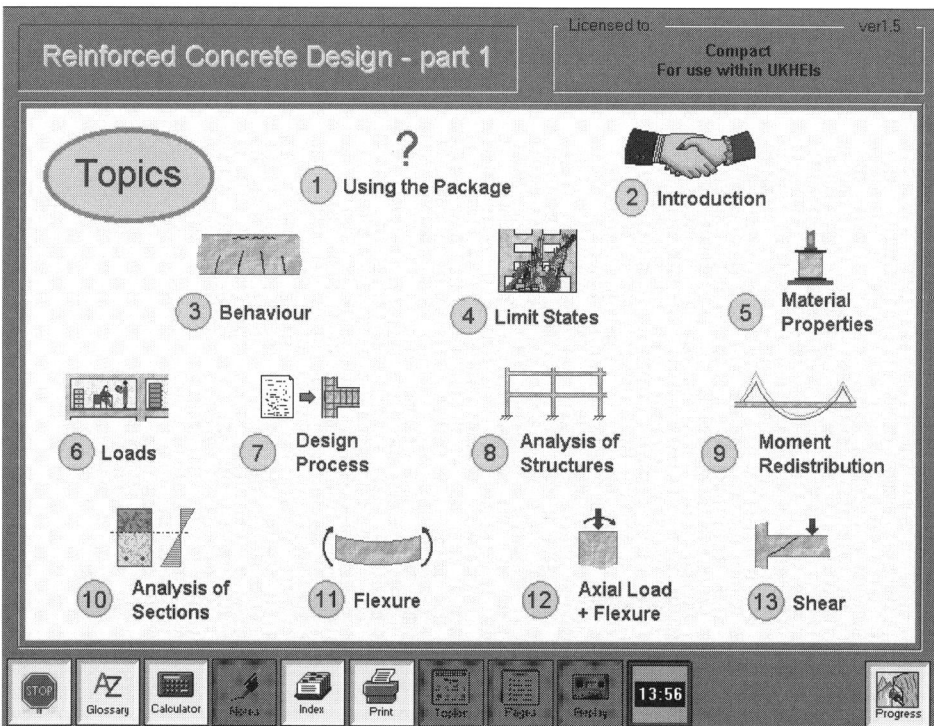

4.1.2 Design of Reinforced Concrete Structures

- Introduction;
- Serviceability;
- Design Details;
- Buckling;
- Torsion;
- Slabs;
- Punching Shear;
- Strut & Tie Models; and
- Shear Walls.

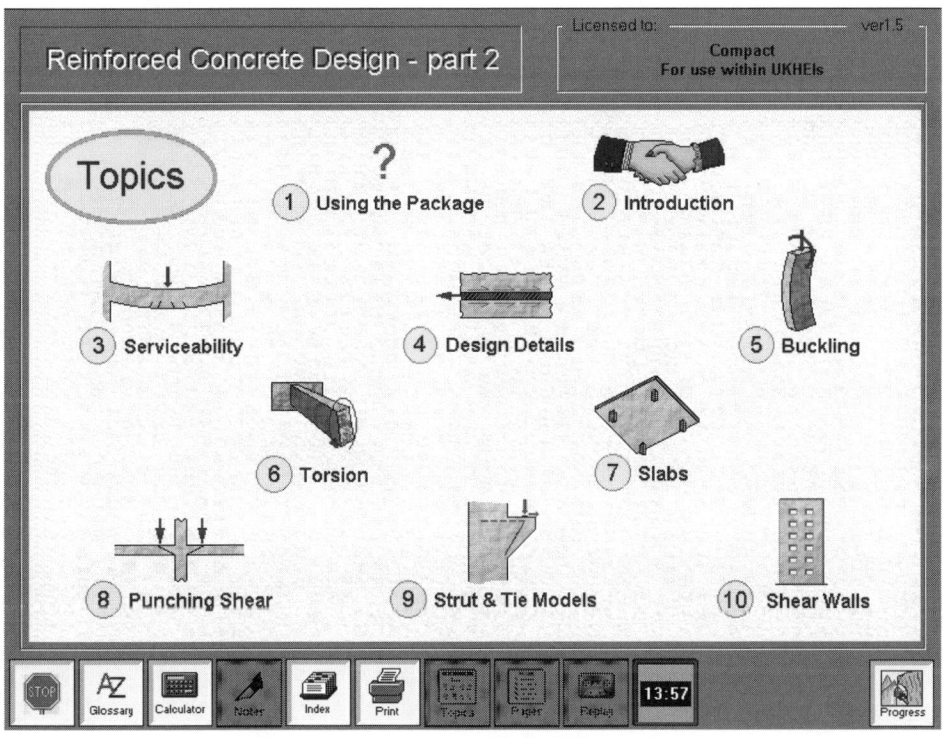

4.1.3 Buildability

- Introduction;
- Evaluation of Buildability;
- Criteria for Buildability;
- General Considerations
- Construction Techniques;
- Construction Processes; and
- Summary/Conclusion.

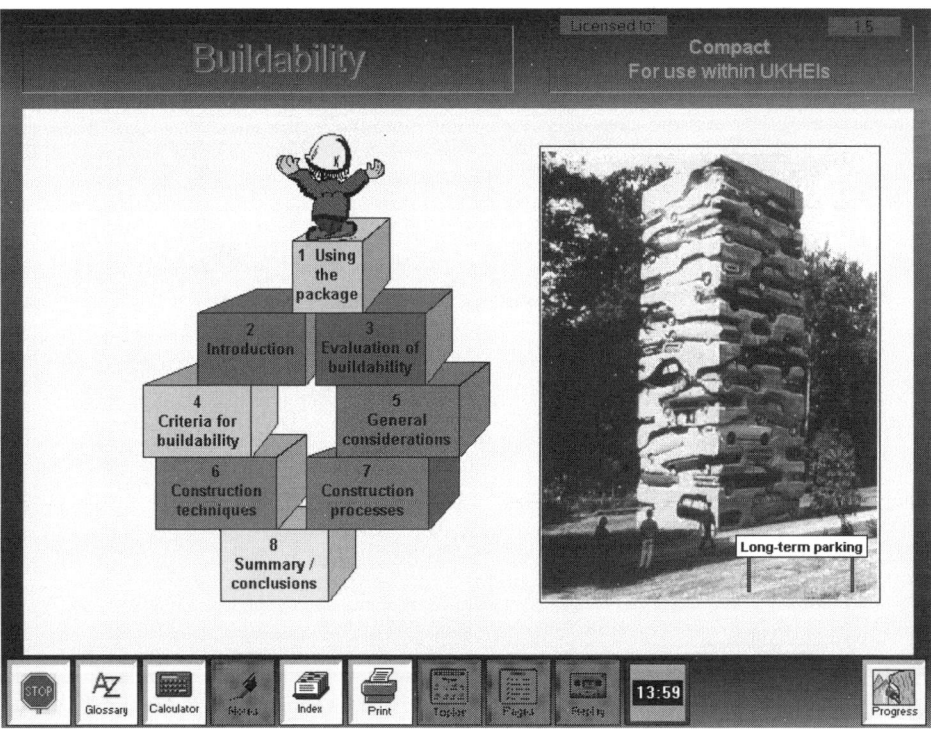

Module 1: Unit 4

4.1.4 Conceptual Design of Concrete Structures

- Introduction;
- The Design Process;
- Elements;
- Stability;
- Structural Concepts; and
- Designing a Tall Building.

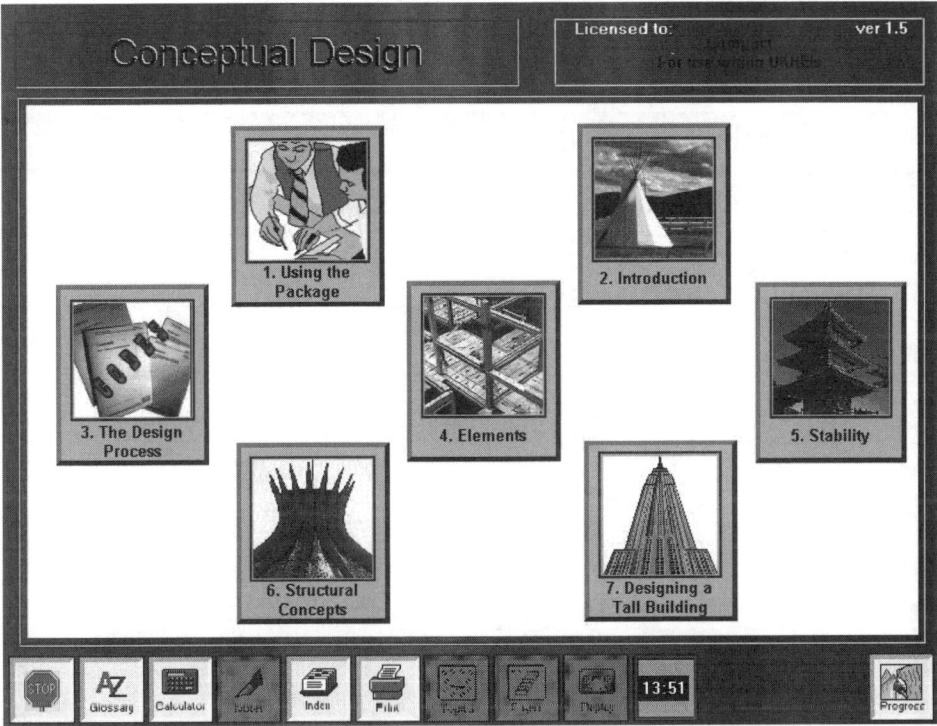

4.1.5 Concrete as a Material (including Mix Design)

- Introduction;
- Cement;
- Aggregates;
- Fresh Concrete (Workability);
- Fresh Concrete (Other Topics);
- Hardened Concrete (Strength);
- Hardened Concrete (Deformation);
- Durability;
- Admixtures; and
- Mix Design.

4.1.6 Concrete Bridges

- Introduction;
- Conceptual Design;
- Aesthetics & Economics;
- Loading;
- Analysis;
- Foundations;
- Decks;
- Joints & Bearing;
- Strengthening; and
- Maintenance & Inspection.

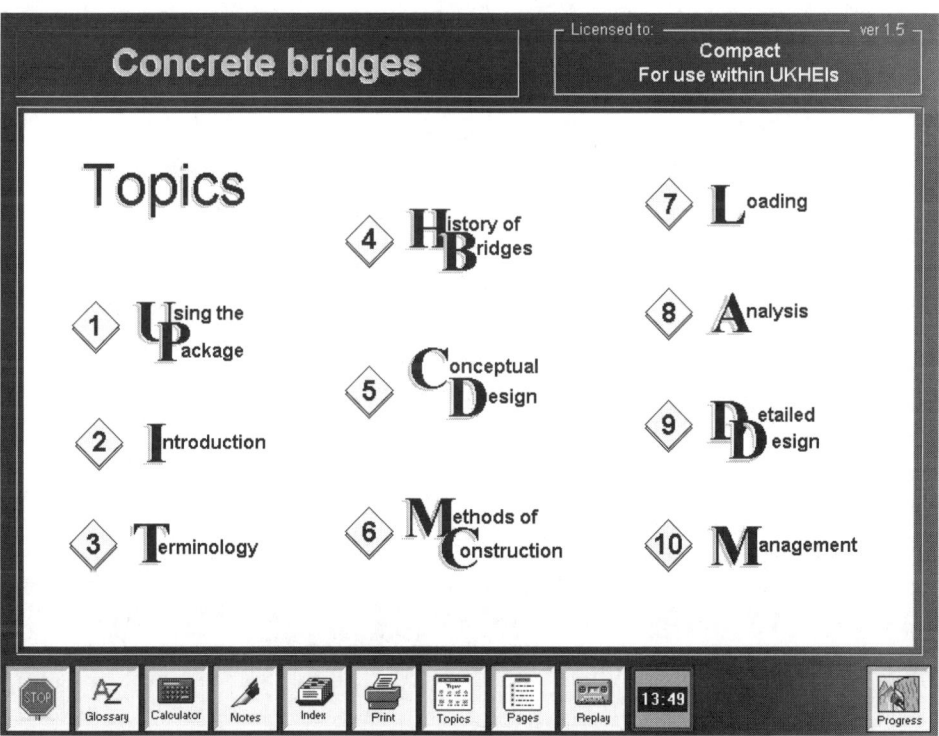

4.1.7 Concrete Site Practice

- Introduction;
- Specification of Concrete;
- Pre-concrete Checks;
- Sampling and Testing;
- Transporting Concrete;
- Placing & Compaction;
- Curing; and
- Post-concrete Inspection.

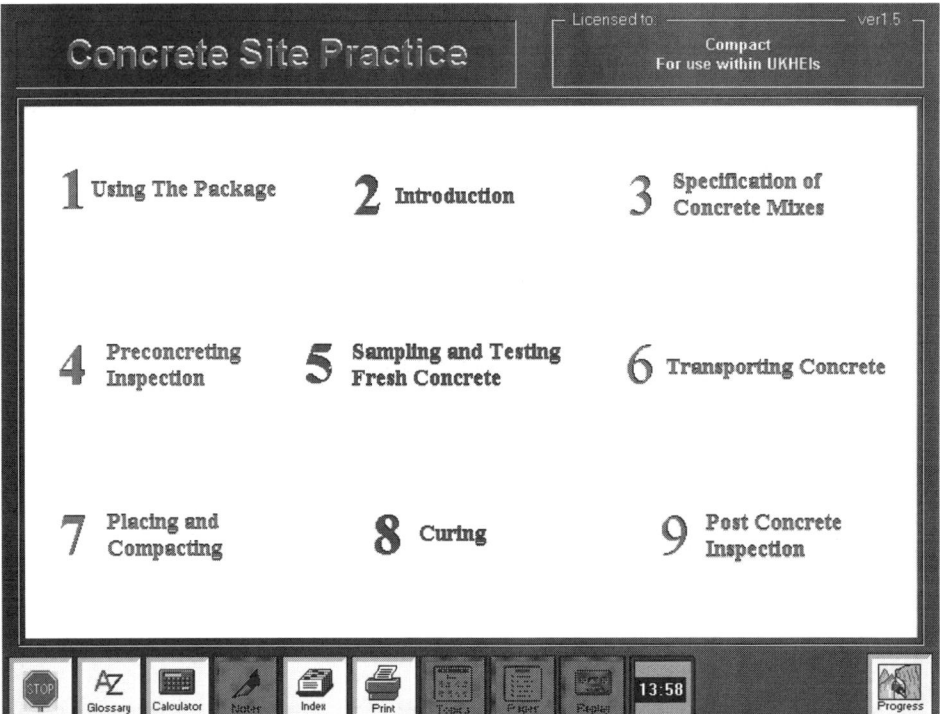

4.1.8 Drawing and Detailing of Concrete Structures

- Introduction;
- Drawings;
- Detailing Data;
- Drawing Layouts;
- Slabs;
- Beams;
- Columns;
- Walls;
- Stairs;
- Foundations; and
- Corbels & Nibs.

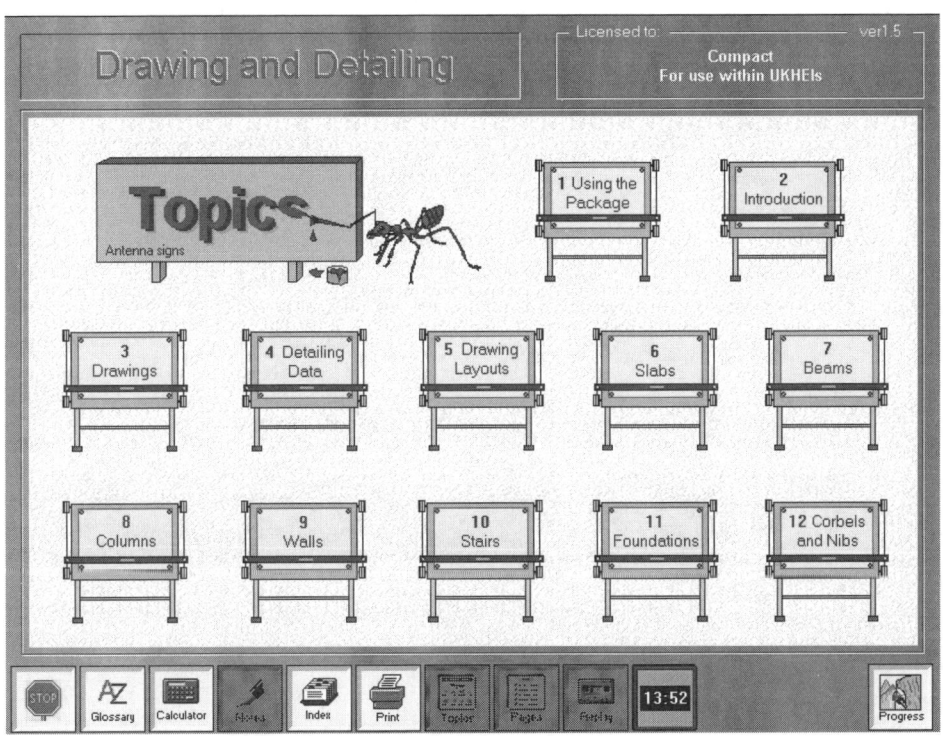

Drawing for Civil Engineering

4.1.9 Foundations and Retaining Walls

- Introduction;
- Pad Footings;
- Strip Footings;
- Raft Foundations;
- Piles & Pilecaps; and
- Retaining Walls.

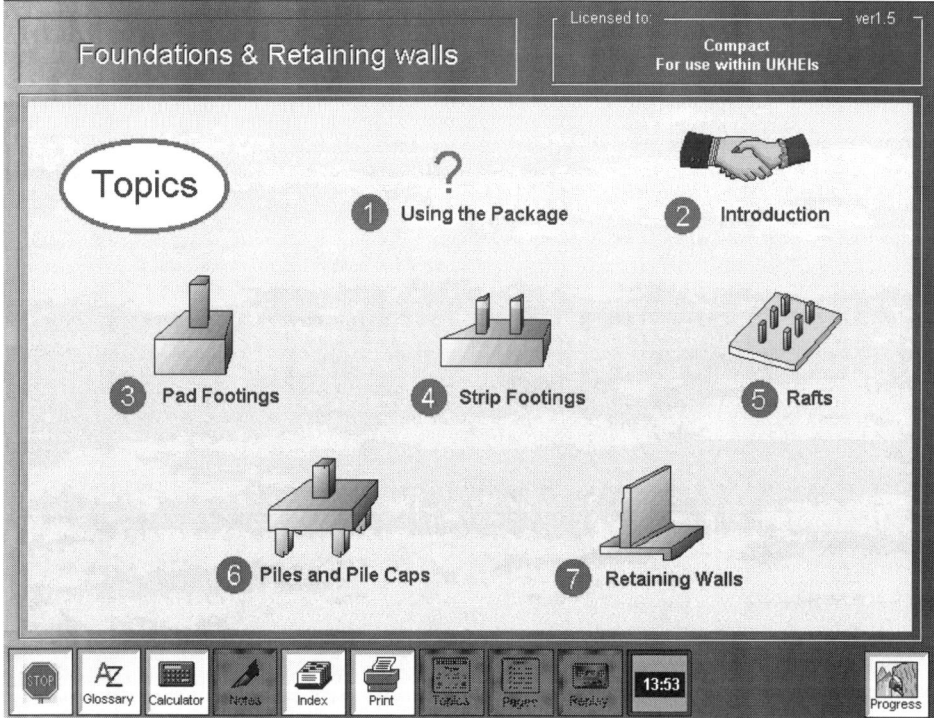

Module 1: Unit 4

4.1.10 Precast Concrete Structures

- Introduction;
- Why Precast;
- Structural Frames;
- Hybrid Construction;
- Components;
- Connections;
- Stabilising Methods;
- Robustness;
- Cladding; and
- Manufacture.

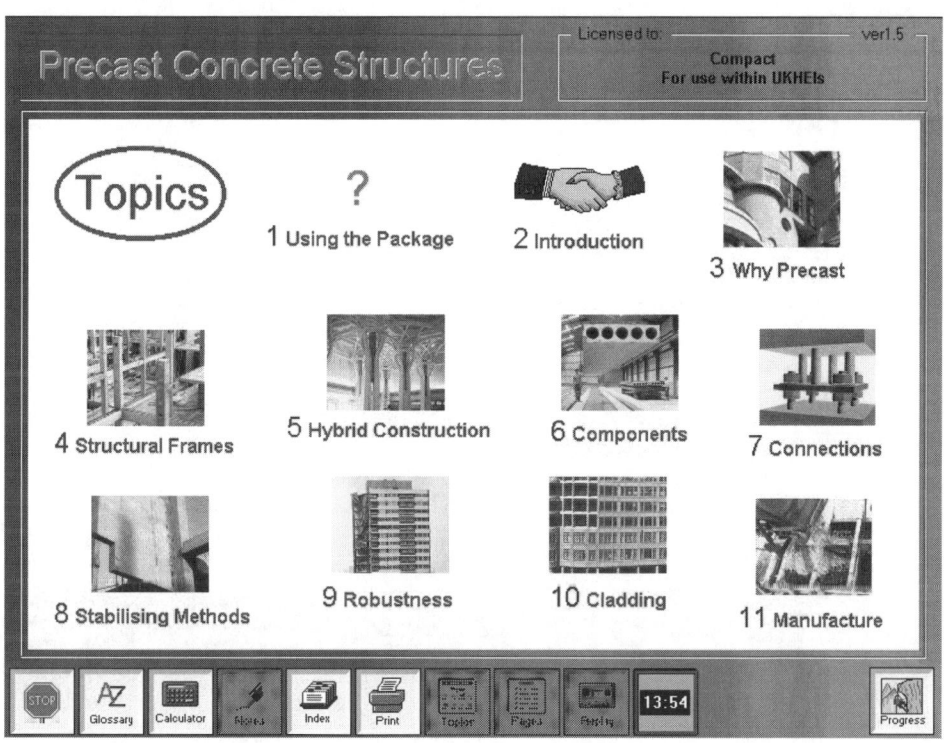

4.1.11 Pre-stressed Concrete Structures

- Introduction;
- Elastic Analysis;
- Load Balancing;
- Losses of Prestress;
- Flexure;
- Shear;
- Project; and
- Continuous Beams.

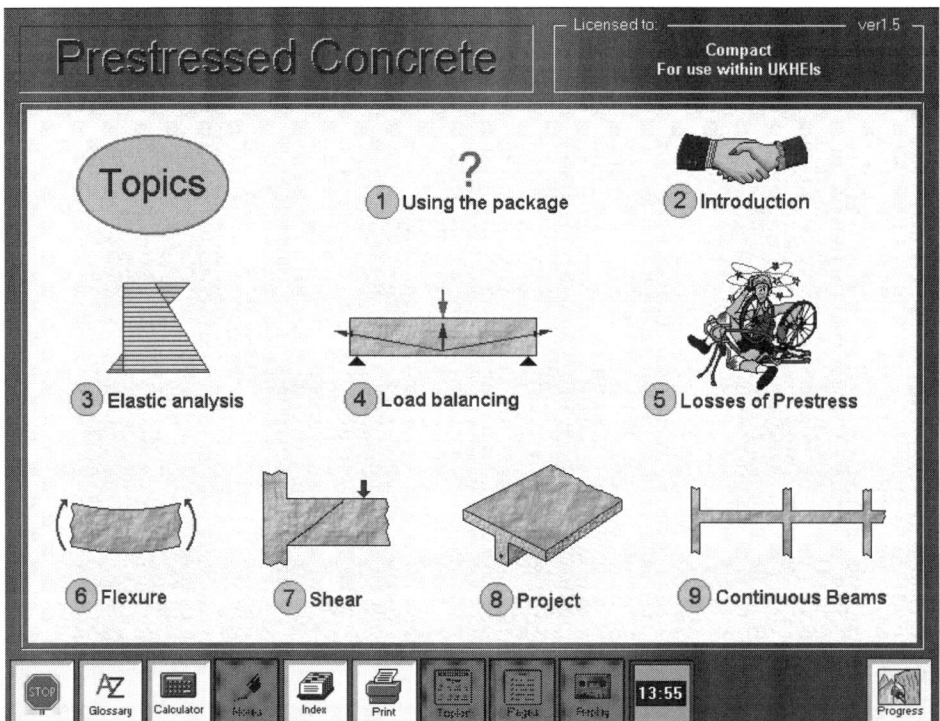

 Not all the above aspects of structural steel construction are covered in the syllabus at this level.
Illustration reproduced with permission from the Southern African Institute of Steel Construction.

4.2 Downloading COMPACT

You will need internet access.
1. Go to the COMPACT website at: http://compact.shef.ac.uk.
2. The webpage should open on the Home Page.
3. The entire COMPACT suite is now available online.
4. Go to 'Study Area'. This area of the website contains resources that students and academics may find useful for their learning and teaching. To use options 1, you will need to have the latest versions of the Shockwave plug-in and the Authorware Web Player installed. To get these, click on the relevant buttons that look like the buttons displayed below.

5. If you have these plug-ins installed, go to 'Free Software' and click on 'Get It' to launch the COMPACT suite online.
6. You will be invited to fill in a form that requires your e-mail address. You will be sent a download key (password) by e-mail. You can then use this key to download the 'Detailing' module. The software should automatically have been saved in c:/compact.
7. Click on DRAW Applications.
8. Click on the appropriate topics, for example:

Topic 1: Using the package
This topic explains how to navigate through the package. There are several different ways of moving to the next piece of information. You will need to recognise the different types of prompt. Read through the four pages.

Topic 2: Introduction
This topic explains what concrete is and explores its history. It gives you background information and enables you to practise using the package so that you are prepared to go into detail later in this unit.

module 2
Structural steelwork

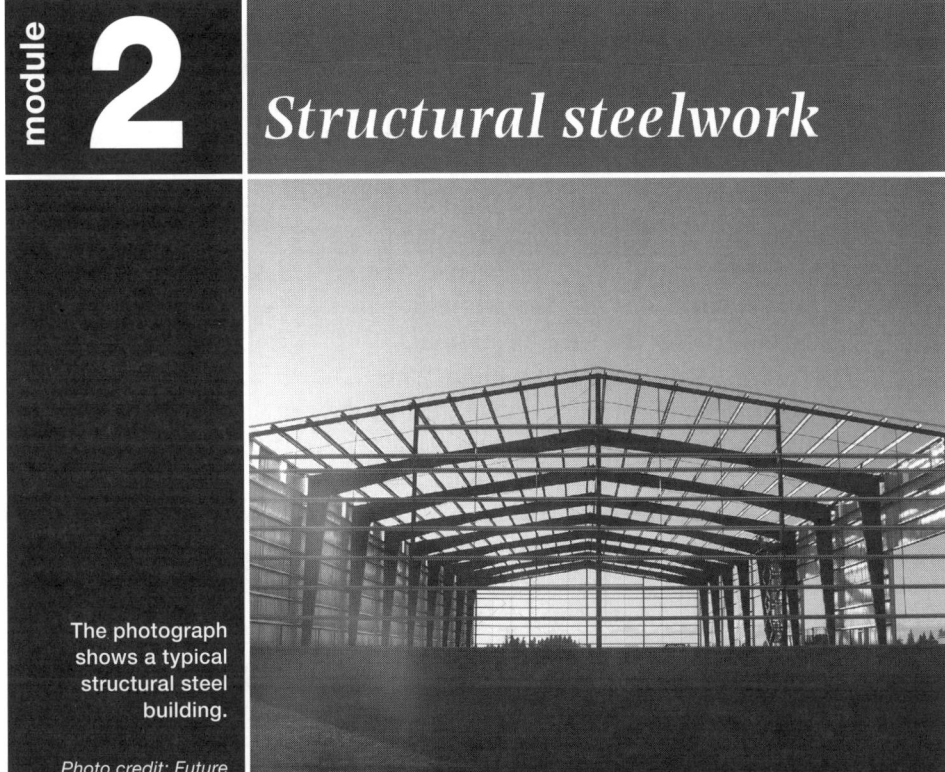

The photograph shows a typical structural steel building.

Photo credit: Future Steel Buildings

Requirements for detailing structural steel drawings

Module outcomes

After studying this module, you should be able to use the different steel tables and codes to produce accurate, neat and correctly dimensioned drawings of:
- typical base-to-column connections showing holding down bolts;
- typical welded and bolted column-to-beam connections;
- typical welded and bolted beam-to-beam connections; and
- different types of welded and bolted roof trusses, lattice girders and portal frames.

Module 2 consists of Units 5, 6, 7, 8 and 9.

Drawing for Civil Engineering

 Read these terms in conjunction with the drawing of a typical structural steel building shown in Fig 5.1 on page 99. The words below are printed in bold the first time they are used in the text.

Axial uplifting forces: Vertical forces that tend to lift in an upward motion.
Cleat: A piece of steel section, bolted or welded to a column, to give footing to a beam or in fixing the cladding.
Electrode: A rod, held between tongs, coated with a flux to form a gaseous shield around the white-hot metal. This provides the right environment for the weld formation.
Grout: A type of mortar for filling gaps.
Gusset: A plate used to connect the members of a truss at the node.
Hand-flame edges: A gas flame is used to cut the edges.
Holing: To pierce or drill through.
Ingot: An oblong piece of steel.
Moment: Rotational motion (that is, a turning effect) = force × distance.
Pitch: The spacing between the centre lines of bolts in the longitudinal direction of a member (beam, girder and the like).

- Compression tends to crush the steel by applying pressing forces to it.
- Tension tends to tear the steel apart by trying to extend it with pulling forces.
- Shearing tends to slice the steel like a guillotine crops a bar in half.
- Bending tends to cause tension to occur on one side of a beam and compression to the other.

Rivet: A type of nail — headless end pressed down when in place — used for holding metal plates together.
Sheared edges: Smooth, mechanically cut edges.
Span: A designed interval between two columns.
Splice plate: A plate used to join two sections (columns and beams) in an overlapping position.
Stanchion: A short column.
Steel mill: A building fitted with machinery for manufacturing steel sections, plates, bars and so on.
Yield stress: The stress beyond which a material becomes plastic. (Remember that you met this term in *Theory of Structures*.)

Module 2

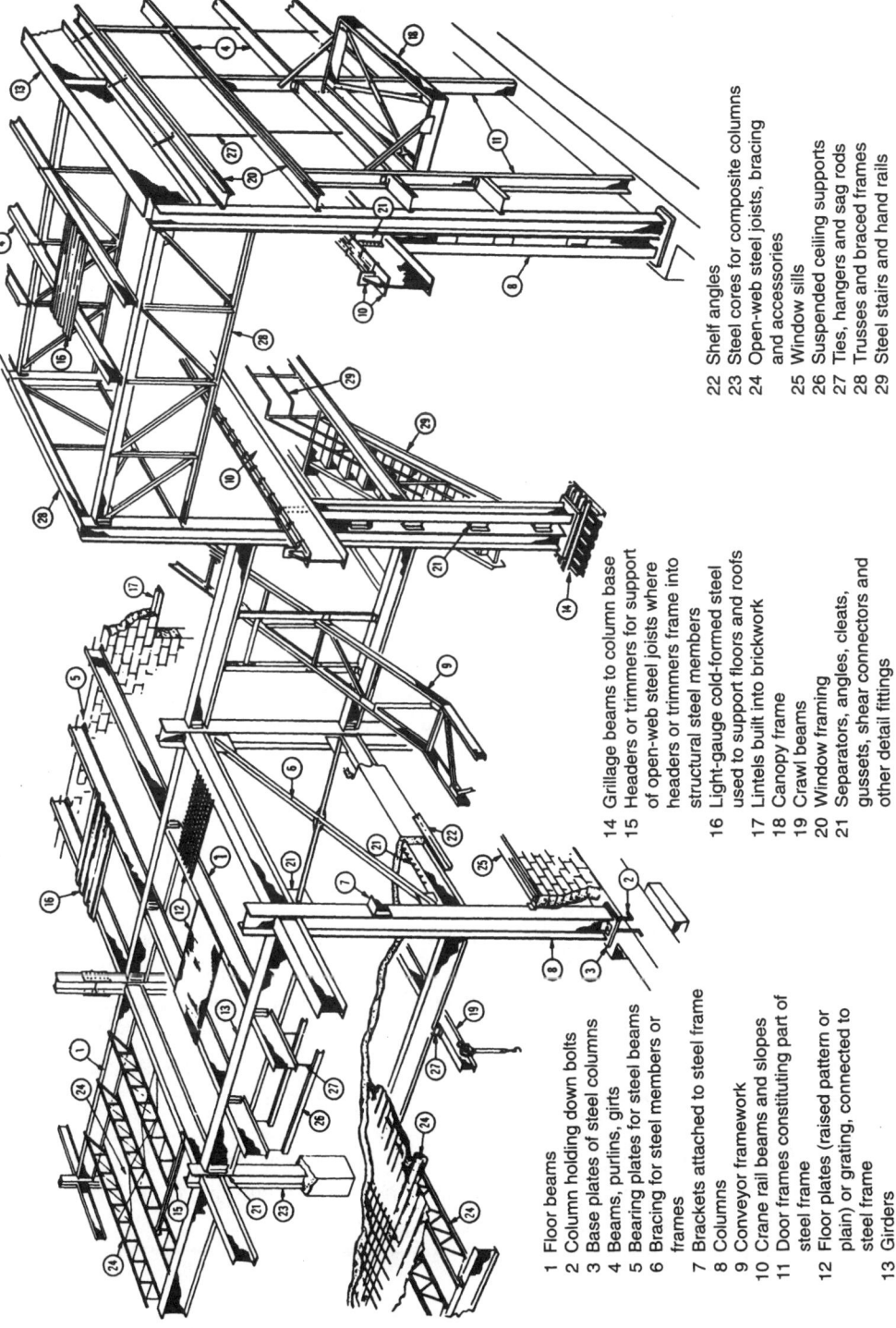

Fig 5.1 Components of a steel frame building

1 Floor beams
2 Column holding down bolts
3 Base plates of steel columns
4 Beams, purlins, girts
5 Bearing plates for steel beams
6 Bracing for steel members or frames
7 Brackets attached to steel frame
8 Columns
9 Conveyor framework
10 Crane rail beams and slopes
11 Door frames constituting part of steel frame
12 Floor plates (raised pattern or plain) or grating, connected to steel frame
13 Girders
14 Grillage beams to column base
15 Headers or trimmers for support of open-web steel joists where headers or trimmers frame into structural steel members
16 Light-gauge cold-formed steel used to support floors and roofs
17 Lintels built into brickwork
18 Canopy frame
19 Crawl beams
20 Window framing
21 Separators, angles, cleats, gussets, shear connectors and other detail fittings
22 Shelf angles
23 Steel cores for composite columns
24 Open-web steel joists, bracing and accessories
25 Window sills
26 Suspended ceiling supports
27 Ties, hangers and sag rods
28 Trusses and braced frames
29 Steel stairs and hand rails

Not all the above aspects of structural steel construction are covered in the syllabus at this level.
Illustration reproduced with permission from the Southern African Institute of Steel Construction.

99

Unit 5 Tables

5.1 Introduction

How many of you have been inside a steel structured building before? It is not surprising if you think that you have not because many people enter a building without knowing what type of building it is. Next time, be on the lookout! Most supermarkets and workshops are, in fact, steel structured buildings. However, *what is structural steel?* Steel is a man-made metal derived from iron; 'structural steel' refers to the various products of a **steel mill**, such as sections, plates and bars. From these, structural members such as beams, girders, columns, struts, ties and hangers are fabricated. (Refer to the drawing of the steel-framed building to see what most of these members look like.)

5.2 Standard steel tables

Structural steel is produced in a number of different strength grades. The grade-designation number gives a direct indication of the **yield stress** of the steel in MPa (that is, mega Pascals). The different forms in which steel are used are I-sections and H-sections, channels, angles, flats, bars, plates, sheets, cold-formed sections and hollow sections. We distinguish between two main classes of sections, namely hot-rolled sections and cold-formed sections.

Can you distinguish between hot-rolled and cold-formed sections?

Hot-rolled sections are produced by passing a large **ingot** of steel, heated to 1 200 °C, through sets of rollers that change the shape of the ingot to suit the section that is required. Standard hot-rolled sections are produced in three grades of steel: 43, 50 and 55. When these numbers are multiplied by 10, they give the maximum tensile strength of the steel. Cold-rolled sections are formed by folding flat-sheet steel into a variety of shapes.

Table 5.1 on pages 102 to 117 shows the dimensions and cross-sectional properties of the range of structural sections that are rolled in South Africa. The standard dimensions of sections are given.

5.3 Bolted connections

You should know what a bolt is, since using bolts is one of the most common methods of joining one component to another in structural steelwork. You learnt about bolts and **rivets** in *S1*. You also need to know that the function of a bolt is not only to attach one member to another, but also to transmit a force from the one member to the other. For example, if bolts are used to connect a beam to a column in a building, the purpose of the bolts is not only to hold the beam in relation to the column, but also to transfer the load to the column.

There are three types of bolts used for connecting structural steelwork:
1. Black bolts — grade 4.6;
2. Black bolts — grade 8.8; and
3. High strength friction bolts.

Let us summarise the important rules regarding the use of bolts (see Fig 5.2 on page 118):
- Minimum **pitch** (s): the distance between the centres of bolts shall not be less than 2,7 times the nominal diameter of the bolt.
- Edge or end distance (a): the minimum edge distance from the centre of any hole to the edge of a plate shall be 1,4 times the nominal diameter of the bolt, except for **sheared edges** or **hand-flame** edges, where it is 1,5 times the nominal diameter of the bolt. Table 5.2 summarises the values for the four commonly used bolt sizes.

Table 5.2 Minimum pitch and edge distance

Bolt diameter (in mm)	12	16	20	24
Edge distance, a: minimum	18	24	30	36
Recommended	25	30	35	40
Pitch, s: minimum	33	44	54	65
Recommended	50	60	70	80

The layout (arrangement) of bolts usually follows a rectangular pattern (as in Fig 5.2a on page 118). When the bolts in alternate rows are offset from those in intermediate rows, they are said to be in staggered pattern, as shown in Fig 5.2b.

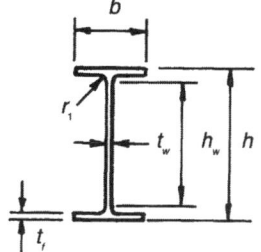

Table 5.1

I-sections (Parallel flange)

Dimensions and properties

Designation mmxmmxkg/m	h mm	b mm	t_w mm	t_f mm	r_1 mm	m kg/m	A 10^3 mm²	I_x 10^6 mm⁴	Z_x 10^3 mm³
IPE-AA 100	97.6	55	3.6	4.5	7	6.72	0.856	1.36	27.8
IPE 100	100	55	4.1	5.7	7	8.1	1.03	1.71	34.2
IPE-AA 120	117	64	3.8	4.8	7	8.36	1.06	2.44	41.7
IPE 120	120	64	4.4	6.3	7	10.4	1.32	3.18	53
IPE-AA 140	136.6	73	3.8	5.2	7	10.1	1.28	4.07	59.7
IPE 140	140	73	4.7	6.9	7	12.9	1.64	5.41	77.3
IPE-AA 160	156.4	82	4	5.6	9	12.3	1.57	6.59	84.3
IPE 160	160	82	5	7.4	9	15.8	2.01	8.69	109
IPE-AA 180	176.4	91	4.3	6.2	9	14.9	1.9	10.2	116
IPE 180	180	91	5.3	8	9	18.8	2.39	13.2	146
IPE-AA 200	196.4	100	4.5	6.7	12	18	2.29	15.3	156
IPE 200	200	100	5.6	8.5	12	22.4	2.85	19.4	194
152x89x16	152.4	88.9	4.6	7.7	7.6	16.1	2.05	8.38	110
178x102x19	177.8	101.6	4.7	7.9	7.6	19	2.42	13.6	153
203x133x25	203.2	133.4	5.8	7.8	7.6	25.3	3.22	23.5	231
203x133x30	206.8	133.8	6.3	9.6	7.6	29.8	3.8	28.9	279
254x146x31	251.5	146.1	6.1	8.6	7.6	31.3	3.99	44.3	352
254x146x37	256	146.4	6.4	10.9	7.6	37.2	4.74	55.5	433
254x146x43	259.6	147.3	7.3	12.7	7.6	43.2	5.5	65.5	505
305x102x25	304.8	101.6	5.8	6.8	7.6	24.5	3.12	43.6	286
305x102x29	308.9	101.9	6.1	8.9	7.6	28.6	3.64	54.4	352
305x102x33	312.7	102.4	6.6	10.8	7.6	32.8	4.18	65	416
305x165x41	303.8	165.1	6.1	10.2	8.9	40.5	5.16	85.5	563
305x165x46	307.1	165.7	6.7	11.8	8.9	46.1	5.88	99.3	647
305x165x54	310.9	166.8	7.7	13.7	8.9	53.5	6.82	117	752

I-sections (Parallel flange)

Dimensions and properties

r_x	I_y	Z_y	r_y	J	C_w	Z_{plx}	Z_{ply}	h/t_f	h_w
mm	10^6 mm^4	10^3mm^3	mm	10^3mm^4	10^9 mm^6	10^3mm^3	10^3mm^3		mm
39.8	0.126	4.57	12.1	7.33	0.272	31.9	7.23	21.7	74.6
40.7	0.159	5.79	12.4	12.1	0.354	39.4	9.15	17.5	74.6
47.9	0.211	6.59	14.1	9.55	0.663	47.6	10.4	24.4	93.4
49	0.277	8.65	14.5	17.4	0.894	60.7	13.6	19	93.4
56.4	0.338	9.27	16.3	12	1.46	67.6	14.5	26.3	112
57.4	0.449	12.3	16.5	24.6	1.99	88.3	19.2	20.3	112
64.8	0.517	12.6	18.1	18.2	2.94	95.2	19.7	27.9	127
65.8	0.683	16.7	18.4	36.2	3.98	124	26.1	21.6	127
73.2	0.781	17.2	20.3	24.9	5.66	130	26.7	28.5	146
74.2	1.01	22.2	20.5	48.1	7.46	166	34.6	22.5	146
81.9	1.12	22.4	22.1	38.7	10.1	176	35	29.3	159
82.6	1.42	28.5	22.4	70.2	13.1	221	44.6	23.5	159
64	0.904	20.3	21	36.1	4.73	124	31.4	19.8	122
74.9	1.38	27.2	23.9	43.8	9.98	171	41.9	22.5	147
85.4	3.09	46.3	31	62.1	29.5	259	71.2	26.1	172
87.2	3.84	57.4	31.8	103	37.3	313	88	21.5	172
105	4.48	61.3	33.5	88.2	66	395	94.2	29.2	219
108	5.71	78	34.7	155	85.7	485	119	23.5	219
109	6.77	92	35.1	242	103	568	141	20.4	219
118	1.19	23.5	19.6	48	26.5	336	37.8	44.8	276
122	1.58	30.9	20.8	78.3	35.5	408	49.2	34.7	276
125	1.94	37.9	21.5	123	44.2	481	60	29	276
129	7.66	92.8	38.5	149	165	626	142	29.8	266
130	8.96	108	39	223	195	722	166	26	266
131	10.6	127	39.4	345	234	843	195	22.7	266

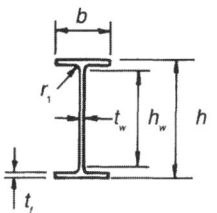

Table 5.1 (continued)

I-sections (Parallel flange)

Dimensions and properties

Designation mmxmmxkg/m	h mm	b mm	t_w mm	t_f mm	r_1 mm	m kg/m	A 10^3 mm^2	I_x 10^6 mm^4	Z_x 10^3 mm^3
356x171x45	352	171	6.9	9.7	10.2	44.8	5.7	121	686
356x171x51	355.6	171.5	7.3	11.5	10.2	50.7	6.46	142	796
356x171x57	358.6	172.1	8	13	10.2	56.7	7.22	161	896
356x171x67	364	173.2	9.1	15.7	10.2	67.2	8.55	195	1070
406x140x39	397.3	141.8	6.3	8.6	10.2	38.6	4.92	124	625
406x140x46	402.3	142.4	6.9	11.2	10.2	46.3	5.9	157	779
406x178x54	402.6	177.6	7.6	10.9	10.2	53.8	6.86	187	927
406x178x60	406.4	177.8	7.8	12.8	10.2	59.7	7.61	215	1060
406x178x67	409.4	178.8	8.8	14.3	10.2	67.1	8.55	243	1190
406x178x75	412.8	179.7	9.7	16	10.2	74.8	9.53	274	1330
457x191x67	453.6	189.9	8.5	12.7	10.2	67.1	8.55	294	1300
457x191x75	457.2	190.5	9.1	14.5	10.2	74.7	9.51	334	1460
457x191x82	460.2	191.3	9.9	16	10.2	82	10.5	371	1610
457x191x90	463.6	192	10.6	17.7	10.2	89.7	11.4	411	1770
457x191x98	467.6	192.8	11.4	19.6	10.2	98.4	12.5	458	1960
533x210x82	528.3	208.7	9.6	13.2	12.7	82.2	10.5	475	1800
533x210x93	533.1	209.3	10.2	15.6	12.7	92.5	11.8	553	2080
533x210x101	536.7	210.1	10.9	17.4	12.7	101	12.9	616	2300
533x210x109	539.5	210.7	11.6	18.8	12.7	109	13.9	668	2480
533x210x122	544.6	211.9	12.8	21.3	12.7	122	15.6	762	2800
533x210x138	549	214	14.7	23.6	12.7	138	17.6	861	3140

I-sections (Parallel flange)

Dimensions and properties

r_x	I_y	Z_y	r_y	J	C_w	Z_{plx}	Z_{ply}	h/t_f	h_w
mm	10^6 mm^4	10^3mm^3	mm	10^3mm^4	10^9 mm^6	10^3mm^3	10^3mm^3		mm
146	8.1	94.7	37.7	160	237	773	146	36.3	312
148	9.68	113	38.7	238	287	895	174	30.9	312
149	11.1	129	39.1	334	330	1010	198	27.6	312
151	13.6	157	39.9	560	413	1210	243	23.2	312
159	4.1	57.8	28.9	108	155	718	90.7	46.2	360
163	5.4	75.9	30.3	194	207	889	119	35.9	359
165	10.2	115	38.6	233	391	1050	178	36.9	360
168	12	135	39.7	332	465	1200	209	31.7	360
169	13.6	153	39.9	465	533	1350	237	28.6	360
170	15.5	173	40.3	642	610	1510	268	25.8	360
185	14.5	153	41.2	376	706	1470	237	35.7	408
187	16.7	176	42	527	820	1660	273	31.5	408
188	18.7	196	42.3	699	923	1830	304	28.8	408
190	20.9	218	42.8	921	1040	2020	339	26.2	408
191	23.5	243	43.3	1220	1180	2230	379	23.9	408
213	20	192	43.8	527	1330	2060	300	40	476
217	23.9	228	45	772	1600	2370	356	34.2	476
218	27	257	45.7	1030	1820	2620	400	30.8	476
219	29.4	279	46	1280	1990	2830	435	28.7	476
221	33.9	320	46.6	1810	2320	3200	500	25.6	477
221	38.7	362	46.9	2540	2670	3610	569	23.3	476

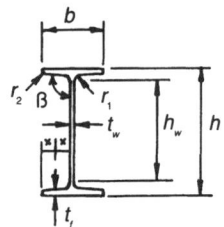

Table 5.1 (continued)

I-sections (Taper flange)

Dimensions and properties

Designation	h	b	t_w	t_f	r_1	r_2	β	m	A
mmxmmxkg/m	mm	mm	mm	mm	mm	mm	deg	kg/m	10^3 mm²
127x76x13	127	76.2	4.5	7.6	7.9	2.4	95	13.4	1.7
152x89x17	152.4	88.9	4.9	8.3	7.9	2.4	95	17.1	2.18
178x102x22	177.8	101.6	5.3	9	9.4	3.2	95	21.4	2.73
203x102x25	203.2	101.6	5.8	10.4	9.4	3.2	95	25.3	3.23
203x152x52	203	152	8.9	16.5	15.5	7.6	98	52.1	6.64

I-sections (Taper flange)

Dimensions and properties

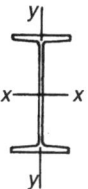

I_y	Z_y	r_x	I_y	Z_y	ry	J	C_w	Z_{plx}	Z_{ply}	h/t_f	h_w
10^6 mm^4	10^3mm^3	mm	10^6 mm^4	10^3mm^3	mm	10^3mm^4	10^9 mm^6	10^3mm^3	10^3mm^3		mm
4.76	74.9	52.9	0.502	13.2	17.2	37.6	1.79	85.2	21.3	16.7	94.2
8.84	116	63.7	0.865	19.5	19.9	53.9	4.49	131	31.5	18.4	118
15.1	170	74.4	1.38	27.3	22.5	81.5	9.86	192	44.2	19.8	138
23	226	84.4	1.63	32.1	22.5	117	15.2	257	51.9	19.5	161
47.8	471	84.8	8.1	107	34.9	667	70.4	539	175	12.3	133

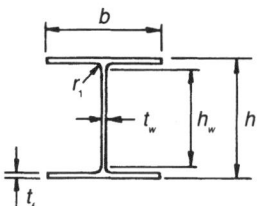

Table 5.1 (continued)

H-sections (Parallel flange)

Dimensions and properties

Designation mmxmmxkg/m	h mm	b mm	t_w mm	t_f mm	r_1 mm	m kg/m	A 10^3 mm^2	I_x 10^6 mm^4	Z_x 10^3 mm^3
152x152x23	152.4	152.4	6.1	6.8	7.6	23.3	2.97	12.6	165
152x152x30	157.5	152.9	6.6	9.4	7.6	30.1	3.84	17.5	222
152x152x37	161.8	154.4	8.1	11.5	7.6	37.1	4.73	22.1	274
203x203x46	203.2	203.2	7.3	11	10.2	46.2	5.88	45.6	449
203x203x52	206.2	203.9	8	12.5	10.2	52.1	6.64	52.5	510
203x203x60	209.6	205.2	9.3	14.2	10.2	59.7	7.6	61	582
203x203x71	215.9	206.2	10.3	17.3	10.2	71.4	9.09	76.3	707
203x203x86	222.3	208.8	13	20.5	10.2	86.4	11	94.6	851
254x254x73	254.2	254	8.6	14.2	12.7	73	9.29	114	896
254x254x89	260.4	255.9	10.5	17.3	12.7	89.2	11.4	143	1100
254x254x107	266.7	258.3	13	20.5	12.7	107	13.7	175	1310
254x254x132	276.4	261	15.6	25.1	12.7	132	16.8	224	1620
254x254x167	289.1	264.5	19.2	31.7	12.7	167	21.2	299	2070
305x305x97	307.8	304.8	9.9	15.4	15.2	96.8	12.3	222	1440
305x305x118	314.5	306.8	11.9	18.7	15.2	118	15	276	1760
305x305x137	320.5	308.7	13.8	21.7	15.2	137	17.4	328	2040
305x305x158	327.2	310.6	15.7	25	15.2	158	20.1	387	2360
305x305x198	339.9	314.1	19.2	31.4	15.2	198	25.2	509	2990

H-sections (Parallel flange)

Dimensions and properties

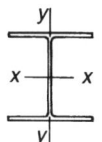

r_x	I_y	Z_y	ry	J	C_w	Z_{plx}	Z_{ply}	h/t_f	h_w
mm	10^6 mm^4	10^3 mm^3	mm	10^3 mm^4	10^9 mm^6	10^3 mm^3	10^3 mm^3		mm
65.1	4.02	52.7	36.8	51.1	21.3	184	80.5	22.4	124
67.5	5.6	73.3	38.2	108	30.7	248	112	16.8	123
68.4	7.06	91.5	38.7	197	39.9	309	140	14.1	124
88.1	15.4	151	51.2	225	142	497	230	18.5	161
89	17.7	173	51.6	322	166	567	263	16.5	161
89.6	20.5	199	51.9	475	195	654	303	14.8	161
91.6	25.3	245	52.8	817	249	801	373	12.5	161
92.7	31.1	298	53.2	1400	317	978	455	10.8	161
111	38.8	306	64.6	578	559	990	463	17.9	200
112	48.3	378	65.2	1040	714	1230	574	15.1	200
113	58.9	456	65.7	1750	893	1480	695	13	200
116	74.5	571	66.6	3180	1180	1860	870	11	201
119	97.9	740	67.9	6340	1620	2420	1130	9.12	200
134	72.7	477	76.8	919	1550	1590	724	20	247
136	90.1	587	77.6	1620	1970	1950	892	16.8	247
137	106	690	78.2	2510	2380	2290	1050	14.8	247
139	125	805	78.9	3810	2850	2670	1230	13.1	247
142	162	1030	80.2	7440	3860	3440	1580	10.8	247

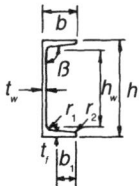

Table 5.1 (continued)

Channels (Parallel flange)

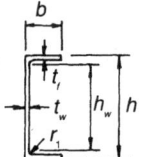

Dimensions and properties

Designation mmxmmxkg/m	h mm	b mm	t_w mm	t_f mm	r_1 mm	m kg/m	A 10^3 mm²	I_x 10^6 mm⁴	Z_x 10^3 mm³
PFC 100x50	100	50	5	8.4	8.4	10.1	1.29	2.05	41.1
PFC 180x70	180	70	7	10.9	10.9	21.1	2.68	13.5	150
PFC 200x75	200	75	7.5	11.4	11.4	24.3	3.09	19.1	191
200x90x30	200	90	7	14	12	29.7	3.79	25.2	252
230x75x26	230	75	6.5	12.5	12	25.7	3.27	27.5	239
230x90x32	230	90	7.5	14	12	32.2	4.1	35.2	306
260x75x28	260	75	7	12	12	27.6	3.51	36.2	278
260x90x35	260	90	8	14	12	34.8	4.44	47.3	364
300x90x41	300	90	9	15.5	12	41.4	5.27	72.2	481
300x100x46	300	100	9	16.5	12	45.3	5.76	81.7	545

Channels (Taper flange)

Designation mmxmmxkg/m	h mm	b mm	t_w mm	t_f mm	r_1 mm	r_2 mm	b_1 mm	á deg	m kg/m	A 10^3 mm²	I_x 10^6 mm⁴
DIN taper flange											
100x50x11	100	50	6	8.5	8.5	4.5	25.6	94.57	10.5	1.34	2.05
120x55x13	120	55	7	9	9	4.5	27.5	94.57	13.3	1.7	3.64
140x60x16	140	60	7	10	10	5	30	94.57	16	2.04	6.05
160x65x19	160	65	7.5	10.5	10.5	5.5	32.5	94.57	18.9	2.4	9.25
180x70x22	180	70	8	11	11	5.5	35	94.57	22	2.8	13.5
200x75x25	200	75	8.5	11.5	11.5	6	37.5	94.57	25.3	3.22	19.1
BS taper flange											
76x38x7	76.2	38.1	5.1	6.8	7.6	2.4	16.5	95	6.72	0.855	0.743
127x64x15	127	63.5	6.4	9.2	10.7	2.4	28.5	95	14.9	1.9	4.83
152x76x18	152.4	76.2	6.4	9	12.2	2.4	34.9	95	17.9	2.28	8.51
178x54x15	177.8	54	5.8	8.3	8.3	3.2	24.1	92	14.6	1.85	8.6

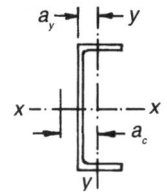
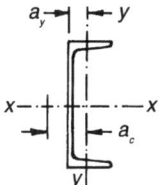

Channels (Parallel flange)

Dimensions and properties

r_x	I_y	Z_y	r_y	J	C_w	Z_{plx}	Z_{ply}	h/t_f	h_w	a_c	a_y
mm	10^6 mm^4	10^3 mm^3	mm	10^3 mm^4	10^9 mm^6	10^3 mm^3	10^3 mm^3		mm	mm	mm
40	0.32	9.79	15.8	24.2	0.49	48.3	17.4	11.9	66.4	34.1	17.3
71	1.27	26.2	21.8	82.3	6.52	177	47.6	16.5	136	43.5	21.5
78.6	1.67	31.8	23.2	104	10.6	225	57.7	17.5	154	45.7	22.5
81.6	3.14	53.4	28.8	183	19.7	291	94.6	14.3	148	63.6	31.2
91.7	1.81	34.8	23.5	118	15.3	278	63.2	18.4	181	47.4	23
92.7	3.34	55	28.6	193	27.9	355	99.1	16.4	178	60	29.2
101	1.85	34.4	23	117	20.4	328	62	21.7	212	43.6	21
103	3.53	56.3	28.2	206	38	425	102	18.6	208	56.5	27.4
117	4.04	63.1	27.7	288	58.2	568	114	19.4	245	53.3	26
119	5.67	81.7	31.4	358	81	636	148	18.2	243	63.1	30.6

Channels (Parallel flange)

Z_x	r_x	I_y	Z_y	r_y	J	C_w	Z_{plx}	Z_{ply}	h/t_f	h_w	a_c	a_y
10^3 mm^3	mm	10^6 mm^4	10^3 mm^3	mm	10^3 mm^4	10^9 mm^6	10^3 mm^3	10^3 mm^3		mm	mm	mm
40.9	39.1	0.29	8.4	14.7	26.8	0.372	48.8	16.1	11.8	64.4	31.7	15.5
60.7	46.3	0.431	11.1	15.9	39.7	0.826	72.7	21.3	13.3	82.1	32.7	16.1
86.4	54.5	0.625	14.7	17.5	54.8	1.64	103	28.3	14	97.9	36.4	17.6
116	62.1	0.851	18.3	18.8	71.2	2.95	138	35.2	15.2	116	38.5	18.4
150	69.6	1.14	22.4	20.1	91.1	5.06	179	43.1	16.4	133	40.6	19.3
191	77.1	1.48	26.9	21.4	115	8.19	228	51.9	17.4	151	42.7	20.1
19.5	29.5	0.107	4.09	11.2	12	0.085	23.5	7.78	11.2	45.8	23.3	11.9
76	50.4	0.672	15.2	18.8	47.2	1.49	89.5	29.3	13.8	84	40.2	19.4
112	61.1	1.14	21	22.3	56.9	3.62	130	41.2	16.9	106	47.8	22.1
96.8	68.1	0.461	11.6	15.8	32.8	2.29	115	21.4	21.4	143	29.3	14.2

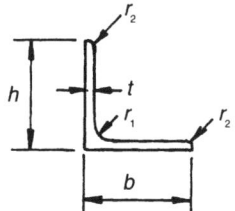

Table 5.1 (continued)

Angles (Equal leg)

Dimensions and properties

Designation mmxmmxkg/m	h mm	b mm	t_w mm	t_f mm	r_1 mm	r_2 mm	b_1 mm	á deg	m kg/m
25x25x3	3.5	2	1.11	0.142	0.008	0.448	7.49	0.013	0.714
25x25x5	3.5	2	1.77	0.226	0.012	0.708	7.3	0.019	1.07
30x30x3	5	2.5	1.36	0.174	0.014	0.649	8.99	0.022	1.05
30x30x5	5	2.5	2.18	0.278	0.022	1.04	8.83	0.034	1.61
40x40x3	6	3	1.85	0.235	0.035	1.18	12.1	0.055	1.93
40x40x4	6	3	2.42	0.308	0.045	1.55	12.1	0.071	2.51
40x40x5	6	3	2.97	0.379	0.054	1.91	12	0.086	3.04
40x40x6	6	3	3.52	0.448	0.063	2.26	11.9	0.1	3.53
45x45x3	7	3.5	2.1	0.268	0.05	1.52	13.7	0.079	2.49
45x45x4	7	3.5	2.74	0.349	0.064	1.97	13.6	0.102	3.2
45x45x5	7	3.5	3.38	0.43	0.078	2.43	13.5	0.124	3.9
45x45x6	7	3.5	4	0.509	0.092	2.88	13.4	0.145	4.56
50x50x3	7	3.5	2.34	0.298	0.07	1.89	15.3	0.11	3.12
50x50x4	7	3.5	3.06	0.389	0.09	2.46	15.2	0.142	4.02
50x50x5	7	3.5	3.77	0.48	0.11	3.05	15.1	0.174	4.92
50x50x6	7	3.5	4.47	0.569	0.128	3.61	15	0.203	5.75
50x50x8	7	3.5	5.82	0.741	0.163	4.68	14.8	0.257	7.27
60x60x4	8	4	3.7	0.471	0.158	3.59	18.3	0.25	5.9
60x60x5	8	4	4.57	0.582	0.194	4.45	18.2	0.307	7.24
60x60x6	8	4	5.42	0.691	0.228	5.29	18.2	0.361	8.52
60x60x8	8	4	7.09	0.903	0.292	6.89	18	0.461	10.9
60x60x10	8	4	8.69	1.11	0.349	8.41	17.8	0.551	13

Module 2: Unit 5

Angles (Equal leg)

Dimensions and properties

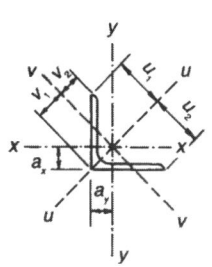

| r_u | I_v | Z_v | r_v | J | a_y |
mm	10^6 mm⁴	10^3 mm³	mm	10^3 mm⁴	mm
9.43	0.003	0.324	4.83	0.476	7.21
9.14	0.005	0.462	4.8	1.98	7.98
11.3	0.006	0.496	5.81	0.635	8.35
11.1	0.009	0.706	5.75	2.58	9.18
15.3	0.014	0.951	7.84	0.882	10.7
15.2	0.019	1.17	7.77	1.92	11.2
15.1	0.023	1.38	7.73	3.56	11.6
14.9	0.027	1.56	7.7	5.92	12
17.2	0.021	1.25	8.88	1.06	11.9
17.1	0.027	1.53	8.76	2.27	12.3
17	0.033	1.8	8.71	4.17	12.8
16.9	0.038	2.05	8.67	6.9	13.2
19.2	0.029	1.58	9.92	1.15	13.1
19.1	0.037	1.94	9.79	2.48	13.6
19	0.046	2.29	9.73	4.58	14
18.9	0.053	2.61	9.68	7.62	14.5
18.6	0.069	3.19	9.63	17	15.2
23	0.066	2.92	11.8	3.07	16
23	0.08	3.45	11.7	5.64	16.4
22.9	0.094	3.96	11.7	9.36	16.9
22.6	0.122	4.86	11.6	21	17.7
22.3	0.148	5.67	11.6	39.2	18.5

113

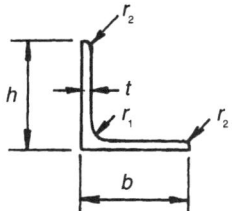

Table 5.1 (continued)

Angles (Equal leg)

Dimensions and properties

Designation hxbxt mm	r_1 mm	r_2 mm	m kg/m	A 10^3 mm²	I_x 10^6 mm⁴	Z_x 10^3 mm³	r_x mm	I_u 10^6 mm⁴	Z_u 10^3 mm³
70x70x6	9	4.5	6.38	0.813	0.369	7.27	21.3	0.585	11.8
70x70x8	9	4.5	8.36	1.06	0.475	9.52	21.1	0.753	15.2
70x70x10	9	4.5	10.3	1.31	0.572	11.7	20.9	0.905	18.3
80x80x6	10	5	7.34	0.935	0.558	9.57	24.4	0.885	15.6
80x80x8	10	5	9.63	1.23	0.722	12.6	24.3	1.15	20.3
80x80x10	10	5	11.9	1.51	0.875	15.4	24.1	1.39	24.5
80x80x12	10	5	14	1.79	1.02	18.2	23.9	1.61	28.4
90x90x6	11	5.5	8.3	1.06	0.803	12.2	27.6	1.27	20
90x90x8	11	5.5	10.9	1.39	1.04	16.1	27.4	1.66	26
90x90x10	11	5.5	13.4	1.71	1.27	19.8	27.2	2.01	31.6
90x90x12	11	5.5	15.9	2.03	1.48	23.3	27	2.34	36.8
100x100x8	12	6	12.2	1.55	1.45	19.9	30.6	2.3	32.5
100x100x10	12	6	15	1.92	1.77	24.6	30.4	2.8	39.6
100x100x12	12	6	17.8	2.27	2.07	29.1	30.2	3.28	46.3
100x100x15	12	6	21.9	2.79	2.49	35.6	29.8	3.93	55.5
120x120x8	13	6.5	14.7	1.87	2.55	29.1	36.9	4.05	47.8
120x120x10	13	6.5	18.2	2.32	3.13	36	36.7	4.97	58.6
120x120x12	13	6.5	21.6	2.75	3.68	42.7	36.5	5.84	68.8
120x120x15	13	6.5	26.6	3.39	4.45	52.4	36.2	7.05	83.1
150x150x10	16	8	23	2.93	6.24	56.9	46.2	9.91	93.4
150x150x12	16	8	27.3	3.48	7.37	67.7	46	11.7	110
150x150x15	16	8	33.8	4.3	8.98	83.5	45.7	14.3	134
150x150x18	16	8	40.1	5.1	10.5	98.7	45.4	16.6	157
200x200x16	18	9	48.5	6.18	23.4	162	61.6	37.2	263
200x200x18	18	9	54.2	6.91	26	181	61.3	41.3	292
200x200x20	18	9	59.9	7.63	28.5	199	61.1	45.3	320
200x200x24	18	9	71.1	9.06	33.3	235	60.6	52.8	374

Angles (Equal leg)

Dimensions and properties

r_u	I_v	Z_v	r_v	J	a_y
mm	10^6 mm^4	10^3 mm^3	mm	10^3 mm^4	mm
26.8	0.153	5.6	13.7	11.2	19.3
26.6	0.197	6.92	13.6	25	20.1
26.3	0.24	8.1	13.5	46.8	20.9
30.8	0.231	7.55	15.7	13	21.7
30.6	0.299	9.37	15.6	29.1	22.6
30.3	0.364	11	15.5	54.5	23.4
30	0.427	12.5	15.5	91.2	24.1
34.7	0.333	9.8	17.8	15	24.1
34.5	0.431	12.2	17.6	33.3	25
34.3	0.526	14.4	17.5	62.4	25.8
34	0.617	16.4	17.4	104	26.6
38.5	0.599	15.5	19.6	37.6	27.4
38.3	0.73	18.3	19.5	70.3	28.2
38	0.857	20.9	19.4	118	29
37.5	1.04	24.4	19.3	221	30.2
46.5	1.05	23.1	23.7	45.4	32.3
46.3	1.29	27.5	23.6	85.1	33.1
46	1.52	31.6	23.5	143	34
45.6	1.85	37.1	23.3	269	35.1
58.2	2.58	45.1	29.7	110	40.3
58	3.03	52	29.5	184	41.2
57.6	3.7	61.6	29.3	347	42.5
57.1	4.35	70.5	29.2	584	43.7
77.6	9.6	123	39.4	564	55.2
77.3	10.7	135	39.3	790	56
77	11.7	146	39.2	1070	56.8
76.4	13.8	167	39	1800	58.4

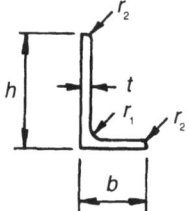

Table 5.1 (continued)

Angles (Unequal leg)

Dimensions and properties

Designation hxbxt mm	r_1 mm	r_2 mm	m kg/m	A 10^3 mm²	I_x 10^6 mm⁴	Z_x 103mm3	r_x mm	I_u 10^6 mm⁴	Z_y 10^3mm³
65x50x6	6	3	5.16	0.658	0.272	6.1	20.3	0.14	3.77
65x50x8	6	3	6.75	0.86	0.348	7.93	20.1	0.177	4.89
75x50x6	7	3.5	5.65	0.719	0.405	8.01	23.7	0.144	3.81
75x50x8	7	3.5	7.39	0.941	0.52	10.4	23.5	0.184	4.95
80x60x6	8	4	6.37	0.811	0.514	9.29	25.2	0.248	5.49
80x60x8	8	4	8.34	1.06	0.663	12.2	25	0.318	7.16
90x65x6	8	4	7.07	0.901	0.734	11.8	28.5	0.323	6.53
90x65x8	8	4	9.29	1.18	0.949	15.5	28.3	0.415	8.54
90x65x10	8	4	11.4	1.46	1.15	19	28.1	0.499	10.4
100x65x8	10	5	9.94	1.27	1.27	18.9	31.6	0.422	8.54
100x65x10	10	5	12.3	1.56	1.54	23.2	31.4	0.51	10.5
100x75x6	10	5	8.04	1.02	1.02	14.7	31.6	0.495	8.67
100x75x8	10	5	10.6	1.35	1.33	19.3	31.4	0.641	11.4
100x75x10	10	5	13	1.66	1.62	23.8	31.2	0.776	14
100x75x12	10	5	15.4	1.97	1.89	28	31	0.902	16.5
125x75x8	11	5.5	12.2	1.55	2.47	29.6	40	0.676	11.6
125x75x10	11	5.5	15	1.91	3.02	36.5	39.7	0.821	14.3
125x75x12	11	5.5	17.8	2.27	3.54	43.2	39.5	0.955	16.9
150x75x10	11	5.5	17	2.16	5.01	51.8	48.1	0.858	14.6
150x75x12	11	5.5	20.2	2.57	5.89	61.4	47.9	0.999	17.2
150x75x15	11	5.5	24.8	3.16	7.13	75.3	47.5	1.2	21
150x90x10	12	6	18.2	2.32	5.33	53.3	48	1.46	21
150x90x12	12	6	21.6	2.75	6.27	63.3	47.7	1.71	24.8
150x90x15	12	6	26.6	3.39	7.61	77.7	47.4	2.05	30.4

Angles (Unequal leg)

Dimensions and properties

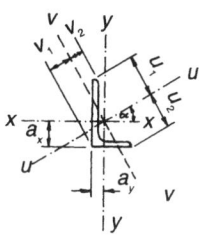

r_y	I_u	Z_u	r_u	I_v	Z_v	rv	J	ax	ay	α
mm	10^6 mm^4	10^3 mm^3	mm	10^3 mm^4	mm	mm	10^3 mm^4	mm	mm	deg
14.6	0.338	7.48	22.7	0.074	3.11	10.6	8.44	20.4	12.9	29.9
14.4	0.43	9.57	22.4	0.096	3.99	10.5	19.1	21.1	13.7	29.7
14.2	0.466	9.1	25.5	0.084	3.17	10.8	9.42	24.4	12.1	23.5
14	0.596	11.7	25.2	0.108	4.1	10.7	21.3	25.2	12.9	23.3
17.5	0.628	11.3	27.8	0.134	4.58	12.9	10.8	24.7	14.8	28.7
17.3	0.808	14.6	27.6	0.173	5.91	12.7	24.4	25.5	15.6	28.6
18.9	0.879	14.1	31.2	0.178	5.44	14.1	11.9	27.9	15.6	27
18.7	1.13	18.3	31	0.23	7.05	13.9	26.9	28.8	16.4	26.9
18.5	1.37	22.2	30.7	0.279	8.58	13.8	50.9	29.6	17.2	26.7
18.3	1.44	21.2	33.7	0.248	7.14	14	29.9	32.7	15.5	22.5
18.1	1.75	25.9	33.5	0.302	8.73	13.9	56.2	33.6	16.3	22.3
22	1.25	17.9	34.9	0.268	7.29	16.2	14.1	30.1	17.9	28.7
21.8	1.62	23.4	34.7	0.346	9.48	16	31.6	31	18.7	28.7
21.6	1.97	28.5	34.5	0.422	11.6	15.9	59.5	31.9	19.5	28.6
21.4	2.3	33.3	34.2	0.495	13.5	15.9	99.8	32.7	20.3	28.4
20.9	2.74	32.5	42.1	0.409	9.75	16.3	36.7	41.4	16.8	19.8
20.7	3.34	39.9	41.8	0.499	12	16.1	69	42.3	17.6	19.7
20.5	3.91	46.9	41.5	0.585	14.1	16.1	116	43.1	18.4	19.5
19.9	5.32	54.4	49.6	0.553	12.3	16	77.4	53.2	16.1	14.6
19.7	6.24	64.2	49.3	0.649	14.6	15.9	130	54.1	16.9	14.5
19.4	7.54	78.4	48.8	0.788	17.9	15.8	246	55.3	18.1	14.2
25.1	5.91	58.5	50.5	0.883	17.5	19.5	83.6	50	20.4	19.8
24.9	6.94	69.1	50.2	1.04	20.7	19.4	141	50.8	21.2	19.7
24.6	8.41	84.2	49.8	1.26	25.3	19.3	266	52.1	22.3	19.5

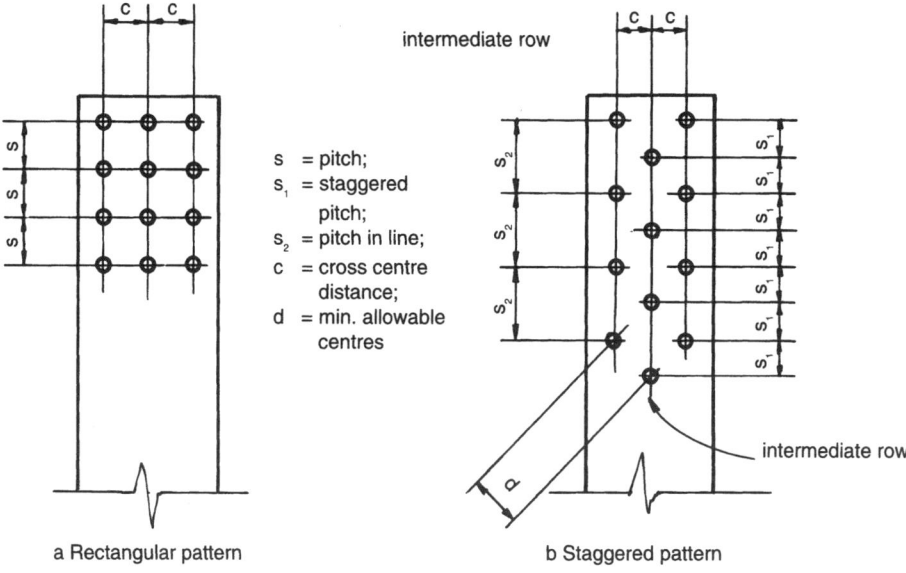

Fig 5.2 Layout of bolts

5.4 Backmark

Holes in the flanges of I- and H-sections are usually placed at a set distance from the web centres (gauge lines or cross-centre lines) or at a set distance from the backs of channels or angles (gauge lines or backmark lines). We refer to this set distance as the gauge distance (see Table 5.3 on page 119).

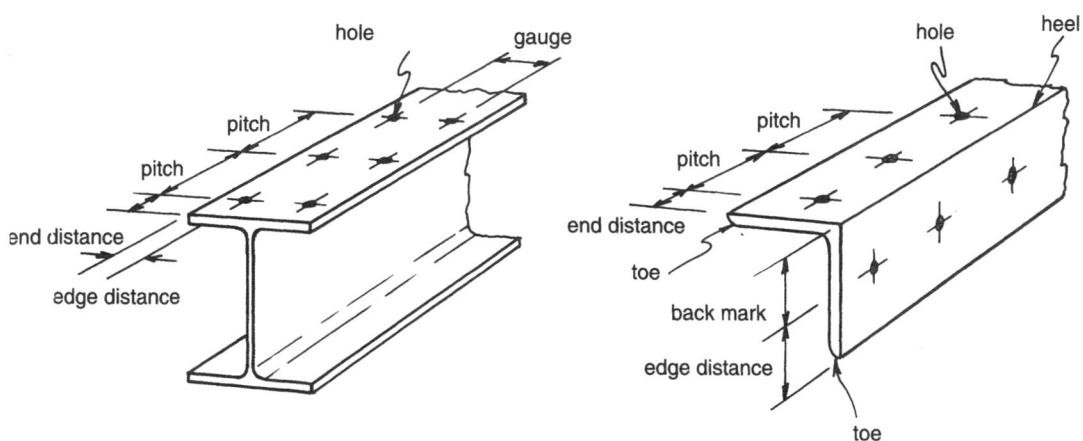

Fig 5.3 Hole positioning and terms

Module 2: Unit 5

Table 5.3 Structural detailing

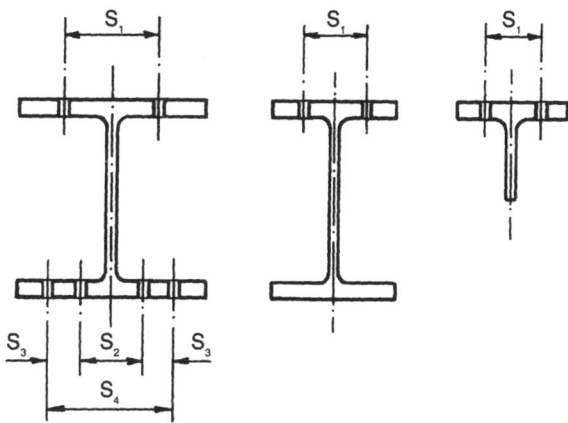

Nominal flange widths (mm)	Spacings (mm)				Recommended dia. of rivet or bolt (mm)	Actual b_{min} (mm)	Nominal flange widths (mm)	Spacing S_1 (mm)	Recommended dia. of rivet or bolt (mm)	Actual b_{min} (mm)
	S_1	S_2	S_3	S_4						
419 to 368	140	140	75	290	24	362	146 to 114	70	20	130
330 to 305	140	120	60	240	24	312	102	54	12	98
330 to 305	140	120	60	240	20	300	89	50		
292 to 203	140				24	212	76	40		
190 to 165	90				24	162	64	34		
152	90				20	150	51	30		

Holes in the flanges of channels as well as the legs of angles are usually placed at a set distance from the heel of the channel or angle. We refer to this set distance as the backmark (see Table 5.4).

 Do you know where the heel and toe of an I-section is? If not, study Fig 5.3. on page 118.

Table 5.4 Spacing of holes in channels and angles

Spacing of holes in channels

Nominal flange width (mm)	S_1 (mm)	Recommended dia. of rivet or bolt (mm)
102	55	24
89	55	20
76	45	20
64	35	16
51	30	10
38	22	

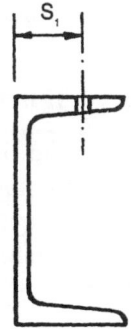

Table 5.4 Spacing of holes in channels and angles (continued)

Spacing of holes in angles

Nominal leg length (mm)	Spacing of holes (mm)						Maximum diameter of bolt or rivet (mm)		
	S_1	S_2	S_3	S_4	S_5	S_6	S_1	S_2 and S_3	S_4, S_5 and S_6
200		75	75	55	55	55		30	20
150		55	55					20	
125		45	50					20	
120		45	50					16	
100	55						24		
90	50						24		
80	45						20		
75	45						20		
70	40						20		
65	35						20		
60	35						16		
50	28						12		
45	25								
40	23								
30	20								
25	15								

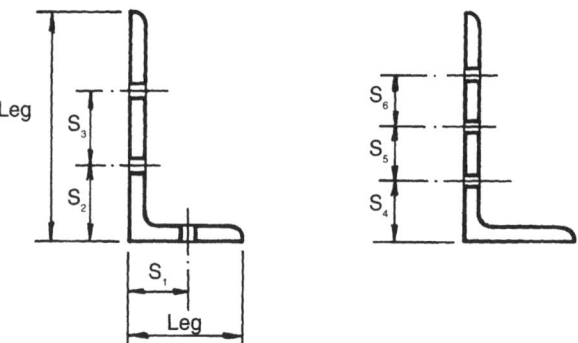

To summarise what we mean by pitch distance, edge or end distances, gauge distances and backmarks, refer to Fig 5.3 on page 118.

5.5 Dimensioning of holes

Fig 5.4 on page 121 shows the two basic methods of dimensioning the position of the holes, from centre to centre. In Fig 5.4a, running dimensions are given from one end of the member (datum). In Fig 5.4b, chain dimensions are given.

Module 2: Unit 5

5.6 Symbols

To make life a little easier, we use standard symbols that the fabricator understands to represent holes, rivets and bolts on drawings. Table 5.5 illustrates these symbols.

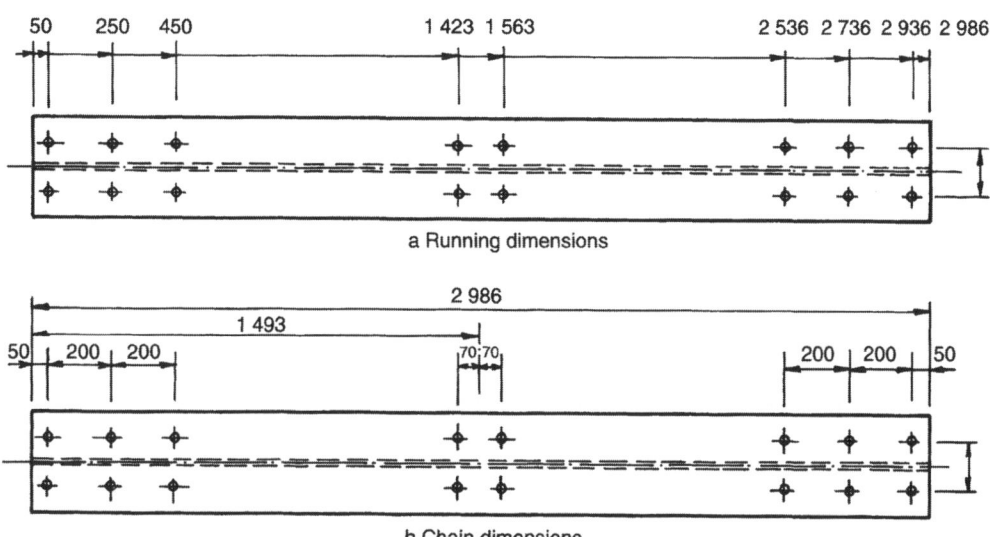

Fig 5.4 Dimensioning of holes along a member

Table 5.5 Symbols for rivets and bolts

Open holes	⊕
Open holes, countersunk near side	⨭
Open holes, countersunk far side	✻
Shop rivets	+
Shop rivets, countersunk near side	⨯
Shop rivets, countersunk far side	✳
Shop bolts	⊕
Shop bolts, countersunk near side	⨭
Shop bolts, countersunk far side	✻
Shop high strength friction grip bolts	⊕
Site high strength friction grip bolts	⊙

Rivets and bolts were covered extensively in *S1*, so read your notes on rivets and bolts again.

5.7 Holding down bolts (HD bolts)

In Unit 2 we look at column bases and columns, and ways of connecting these. But, while we are talking about bolts, let us discuss them in relation to column bases and columns.

Holding down bolts (HD bolts) have two purposes, namely:
1. to position the column base accurately. *Do you remember what we discussed in Unit 2, in Module 1?*
2. to transfer **axial uplifting forces** and bending **moment** from the base of the column into the foundation.

Although HD bolts are set into the concrete by the civil contractor, the steelwork contractor designs their amount, location, diameter and bond length. Usually, every base plate requires a minimum of two bolts and a maximum of eight bolts.

An example of a typical bolt can be seen in Fig 5.5 on page 123. The standard diameters used for HD bolts are 20 mm, 24 mm and 30 mm, and then upwards in 5 mm increments. (Refer to your *S1* notes on diameter again.) The hole sizes in the base plate are made considerably larger than the bolt size, to allow for inaccuracies in bolt setting. Suitable hole sizes for the various bolt diameters are shown in Table 5.6 ('M' means metric).

Table 5.6 Hole diameter

Bolt size	Hole diameter (mm)
M20	26
M24	30
M30	40
M36	46
M42	55
M48	60
M56	70
M64	80
M72	90

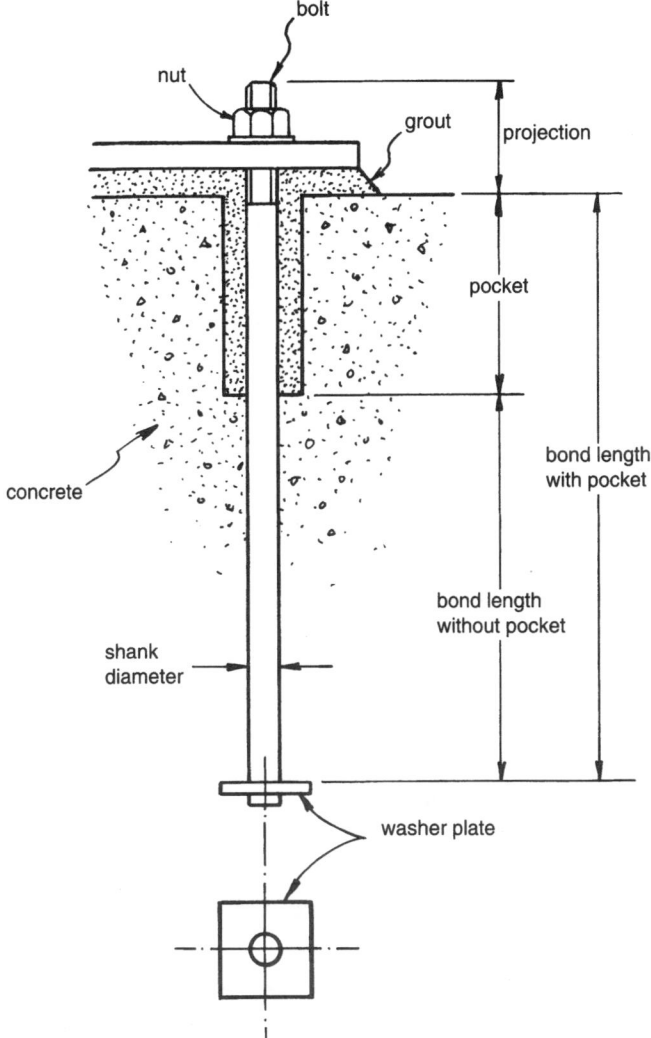

Fig 5.5 Holding down bolt

5.8 Welded connections

What is welding? Welding is the process of joining two pieces of steel together by heating their edges to a molten state so that they fuse together. Additional weld metal is introduced by means of an **electrode**, as part of the process of building up the weld. Welding is an alternative to shop bolting. Welded connections have the advantage of eliminating the weight of the **cleats** and **splice plates**, avoiding strength loss of sections by drilling holes in them, and resulting in a smoother and more easily maintained profile.

5.8.1 Types of welds

Fig 5.6 illustrates the two basic welds that are used, namely fillet welds and butt welds.

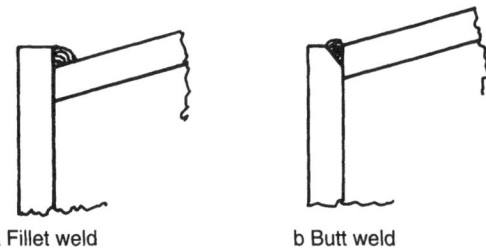

a Fillet weld   b Butt weld

Fig 5.6 Two basic welds

Fillet welds are the most common type of welds and designers prefer to extend one plate in order to produce a fillet weld instead of a butt weld. The reason for this is that butt welds often require the butting edges to be specially prepared, which is time consuming and more costly.

5.8.2 Symbols for welds

Table 5.7 shows the various types of welds. Try to visit a structural steel workshop to see how these types of welding are done. This will help you to understand the different symbols shown in Table 5.7.

Table 5.7 Symbols for welds

Designation	Illustration	Symbol	Designation	Illustration	Symbol
Butt weld between flanged plates (flanges being melted down completely)			Single J-butt weld		ᑭ
Square butt weld		‖	Backing or sealing run		⌣
Single-bevel butt weld		V	Fillet weld		◺
Single V-butt weld		⋁	Plug weld (circular or elongated completely filled)		⊓
Single V-butt weld with broad root face		Y	Spot weld or projection weld a) resistance		○
Single-bevel butt weld with broad root face		Ⴘ	b) arc		○
Single-U butt weld		⋃	Seam weld		⊖

Module 2: Unit 5

5.8.3 Sizes of welds

In situations where you are required to state the size of a weld on a drawing, place the number representing the size of the weld in millimetres (mm) immediately to the left of the weld symbol, as shown in Fig 5.7.

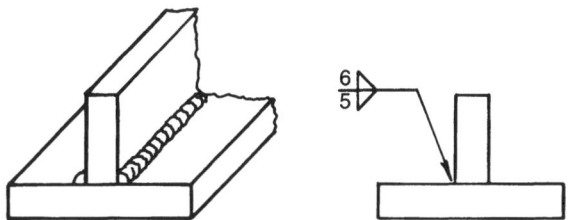

Fig 5.7 Welding both sides (the number indicates the size of the weld)

The symbols for welds are used together with a sloping arrow line that points to the location of the weld. Attached to the arrow line is a reference line that is drawn horizontally. There must be a change of direction between the arrow line and the reference line, as shown in Fig 5.8.

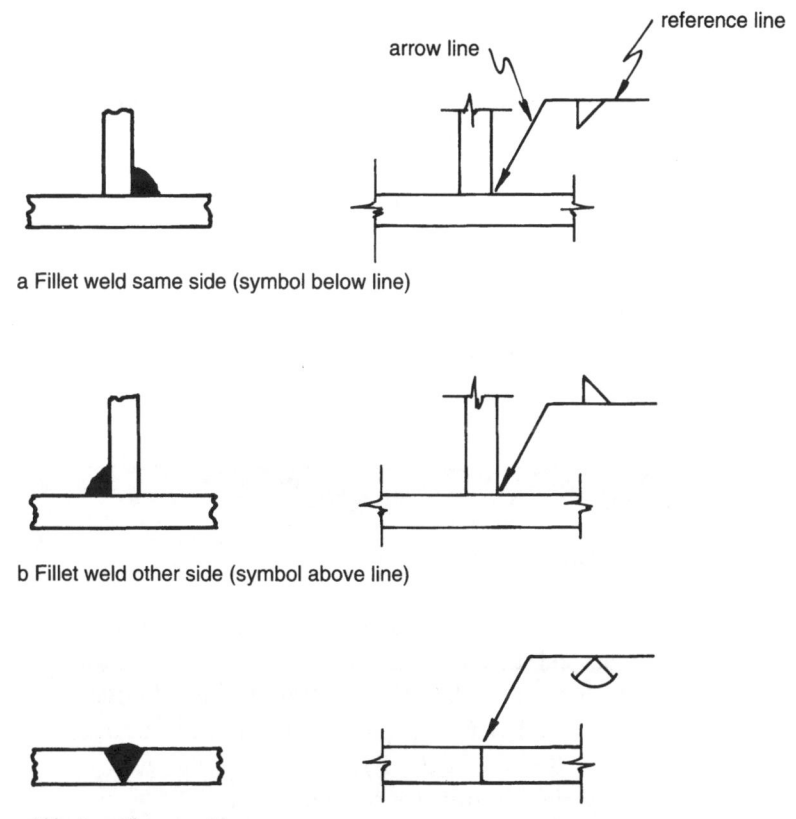

a Fillet weld same side (symbol below line)

b Fillet weld other side (symbol above line)

c V butt weld same side

Fig 5.8 Indicating welds

125

Sometimes a weld is not continuous, in which case the length of a weld and the length of space between welds are each indicated by two consecutive numbers to the right of the symbol, as shown in Fig 5.9 on page 127.

Welding can be indicated on a drawing either by a very thick line where it occurs or by short 45° hatching.

5.9 Summary

1. In this unit, you were introduced to the components of a steel-frame building.
2. You are now familiar with the various tables used in the design and drawing of structural steelwork.
3. You now know that we use two methods to connect the components in structural steelwork, namely bolting and/or welding.

Activity 5.1

(Answers not included)

This activity involves using the tables in this unit to read off the correct values. Consider the following I-section (parallel flange):

$254 \times 146 \times 37$ from Table 5.1 on pages 102 to 117

In the first column, you will find a 254 mm × 146 mm × 37 kg/m I-section is indicated. When you look at the top left-hand corner of page 104, you will find a definition sketch, which indicates the symbols corresponding to the standard dimensions and properties of this I-section.

Activity 5.2

(Answers not included)

Refer to Table 5.1 on pages 102 to 117 and draw, to a suitable scale, the following steel sections:
1. $178 \times 102 \times 19$ kg/m I-section (parallel flange);
2. $203 \times 203 \times 60$ kg/m H-section (parallel flange);
3. $152 \times 89 \times 17$ kg/m I-section (taper flange);
4. 100×50 channel (parallel flange);
5. $70 \times 70 \times 8$ angle (equal leg); and
6. $150 \times 90 \times 15$ angle (unequal leg).

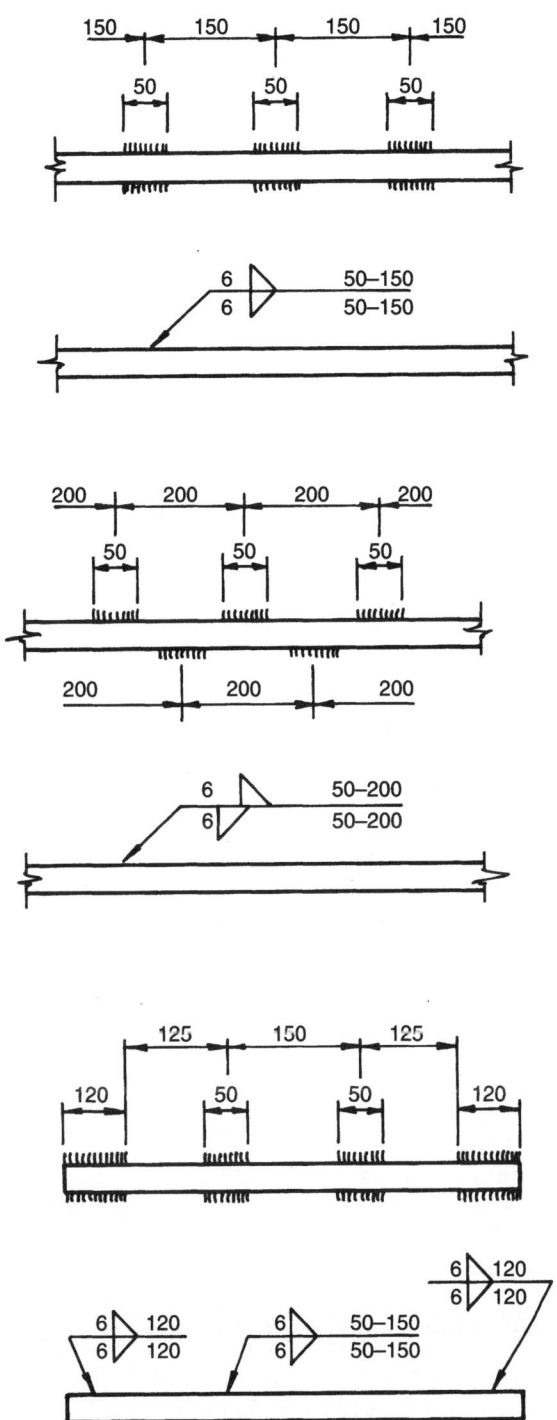

Fig 5.9 Intermittent welding

Self-evaluation

1. What do you understand by the term 'structural steel'?
2. What is meant by the grade of structural steel?
3. How would you describe minimum pitch?
4. What is a backmark?
5. What are the functions of holding down bolts?
6. What is welding?
7. How can you represent welding on a drawing?

SELF-EVALUATION ANSWERS
1. 'Structural steel' refers to the various steel sections produced in a steel mill.
2. The grade is the direct indication of the yield stress of the steel.
3. Minimum pitch is the minimum distance between centres of bolts.
4. A backmark is a set distance from the heel of the section in angles and channels.
5. The functions of HD bolts are:
   - to position the column base; and
   - to transfer axial uplifting forces and moments from the base into the foundation.
6. Welding is the fusion of two pieces of steel by heat.
7. Welding can be shown by:
   - a thick line; or
   - short 45° hatching.

Unit 6 Base-to-column connections

6.1 Bases

In Unit 2 of Module 1 you were introduced to the use of reinforced concrete bases and how to draw them. In this unit, we discuss bases as part of structural steelwork. You should remember that the function of a base is to transmit the vertical loads as well as any possible bending moments into the foundation of the column. We combine the column to the foundation by means of a base plate. The foundation material is usually concrete.

The simplest base plate is required only to transmit the vertical load and keep the column in place. In other words, this base will stop lateral movement. Fig 6.1 illustrates this type of base, which is termed a 'slab base'. Instead of the welding, which Fig 6.1 shows, the base could be bolted to the column by means of angle cleats.

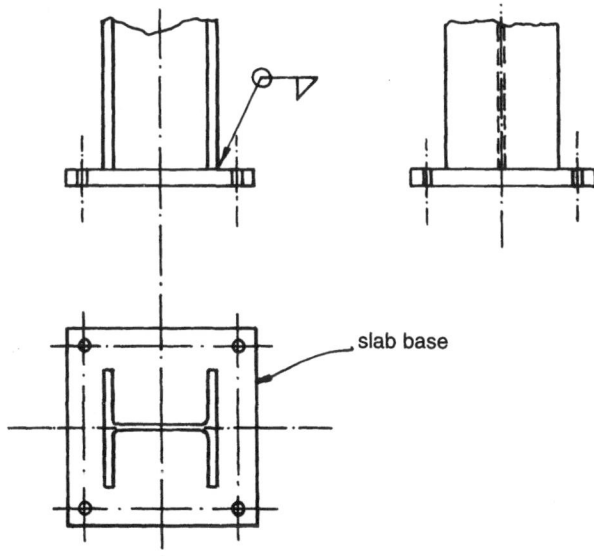

Fig 6.1 Slab base butt-welded to bottom of column

Base plates can also be made to transmit turning forces by stiffening the base plates with **gussets**, as shown in Fig 6.2 on page 130.

As previously indicated, columns transmit generally compressive loads. Therefore, it is important that there is proper contact at the interface between the base plate and the foundation (see Fig 6.3b on page 132). This is because without proper contact, the load will not be transmitted uniformly throughout the base. The base plate is bolted to the foundation to make sure that there is proper contact. *How is proper contact achieved?*

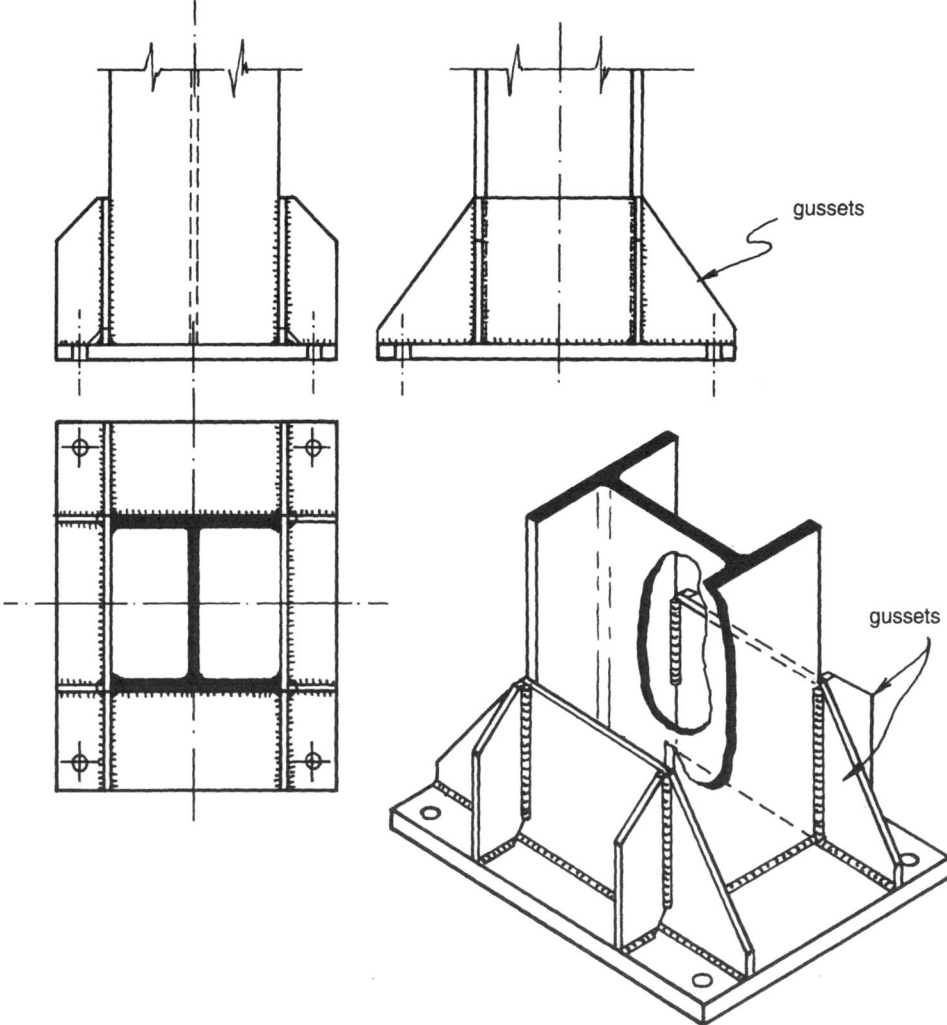

Fig 6.2 Slab base welded to bottom of column and strengthened to resist bending by means of gussets

We leave a space between the underside of the base plate and the top of the concrete foundation. This allows for accurate levelling of the steelwork. We fill the space with **grout**, to ensure a uniform spread of the load from the column onto the concrete surface (see Fig 5.5 on page 123 again).

6.2 Columns

6.2.1 Types of columns

By now you will have observed that many buildings fall into the structural steel category. These buildings vary in size, from one-storey heights to multi-storey heights. This explains why columns vary in size and shape. Fig 6.3 on page 132 shows the four different types of columns that are generally used. Fig 6.3 detail (a) shows a simple single-storey column in a warehouse-type building. Detail (b) shows a column in a portal-frame building; a crane gantry is incorporated into many such buildings. The crane beams are supported on brackets off the column. Detail (c) shows an internal column in a multi-bay workshop built to accommodate heavy overhead cranes. Detail (d) shows an internal column in a multi-storey building and shows how the load is introduced into the column progressively from top to bottom (shown by points A, B and C).

Columns may be detailed vertically or horizontally. Your front view is always the view that looks into the web with the flanges on the left and on the right-hand side (when you detail the column vertically). However, if you choose to detail the column horizontally, your flanges are now positioned at the top and at the bottom of the web, with the base end on the right-hand side of the drawing. Auxiliary views, as indicated in Fig 6.4 on page 133, are normally given to clarify the **holing** arrangement. Fig 6.4 shows a typical column detail.

6.2.2 The grid system

Fig 6.5a on page 134 shows a marking plan for one floor of a simple structure, using a grid system. Note that the grid is lettered A to H from left to right, and 1 to 3 from top to bottom.

Fig 6.5b on page 134 shows a section through a four-storey building on which the floors are lettered alphabetically from the bottom up. If you combine the two figures every member in a structure can be given a reference that uniquely describes its position.

The columns are 'marked' according to the grid intersection at which they are to be erected — A1 to H3. If there is more than one lift of columns (as in Fig 6.5b), their vertical location can be described by the floor code for their lower end. For example, A1-A starts at the ground floor level, continues to the first floor as A1-B, and so on up to A1-E at the fourth floor.

Please note that when you consider Fig 6.5b you must remember that each floor, except the ground floor, has the same layout as Fig 6.5a.

6.2.3 Column splices

Do you know what restricts the length of columns? Do you think it is possible to fabricate a 100m length of steel section at the mill? Even if it could be made, would it be possible to transport such a length?

The solution to this problem is to fabricate only standard, convenient lengths, and then combine them by means of splice plates. The main function of splice plates is to line up the two parts of a column. It is common practice to arrange splices in columns (or **stanchions**) at a

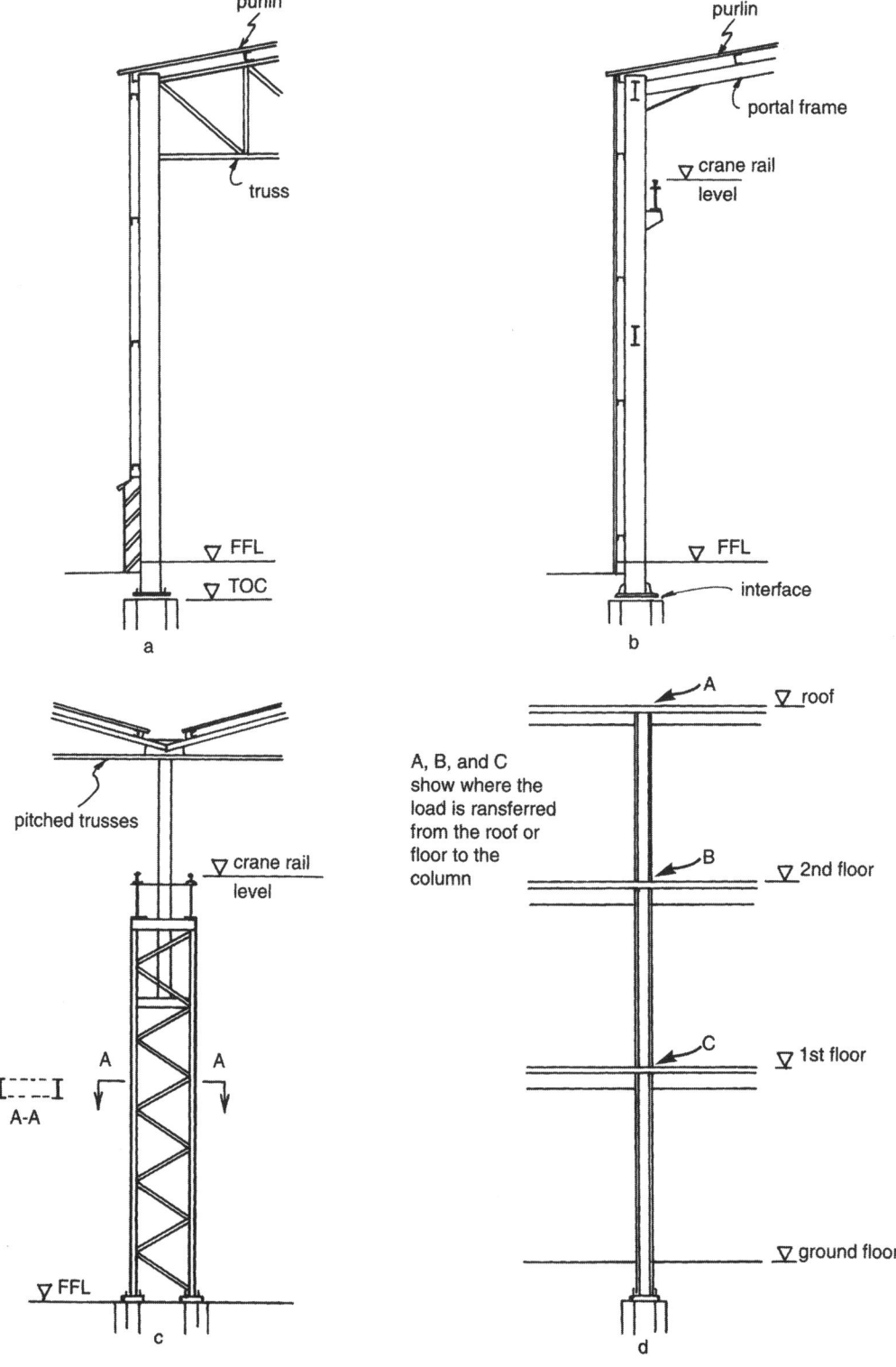

Fig 6.3 Types of columns

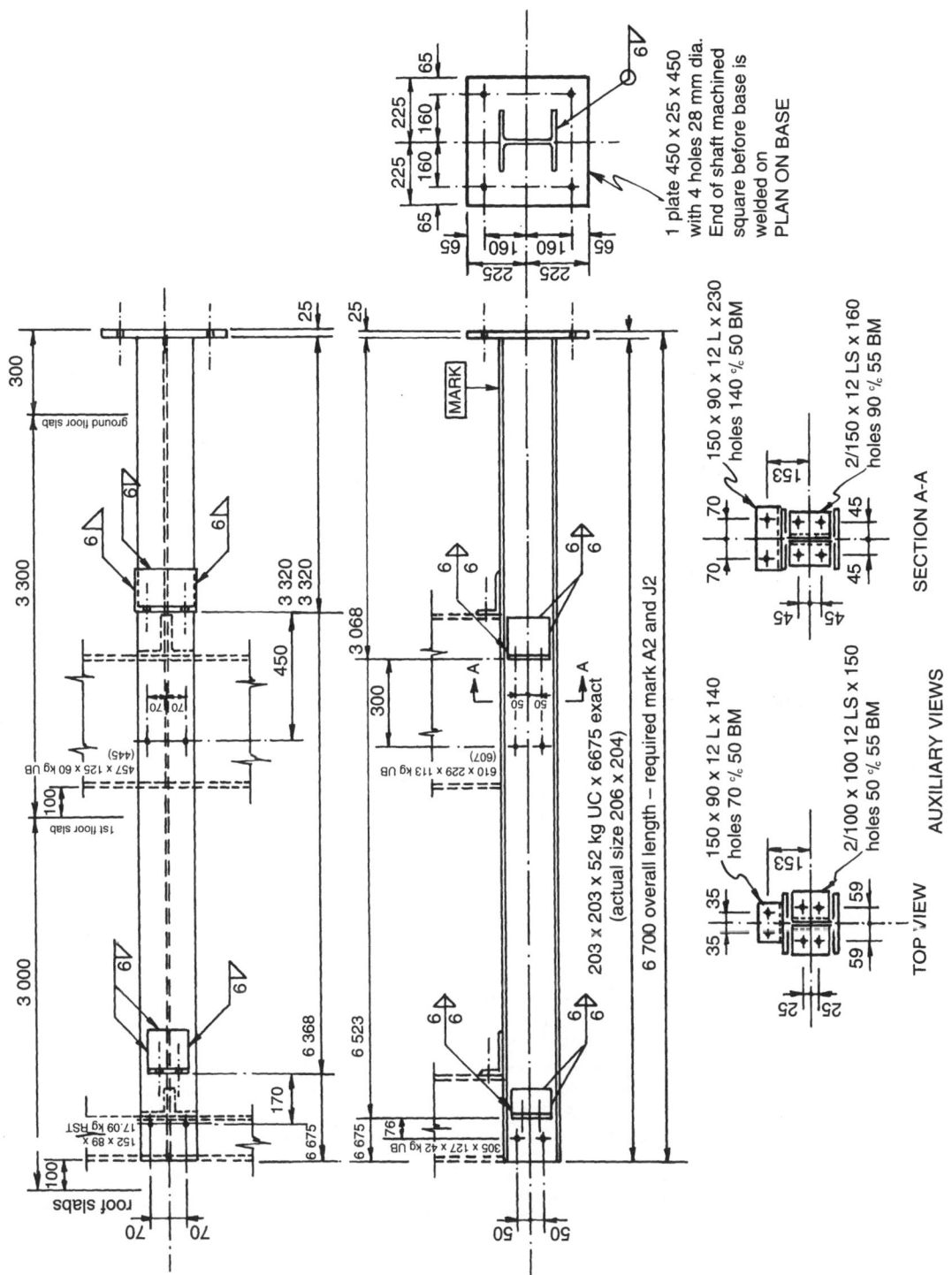

Fig 6.4 Typical column details

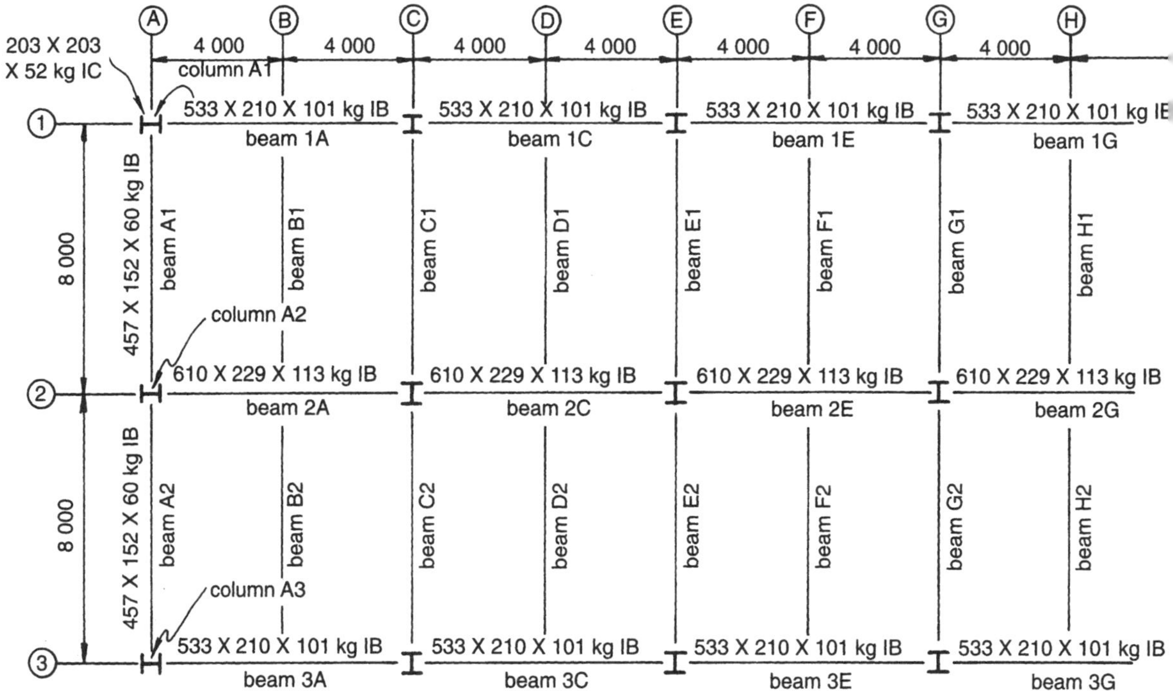

Drawing for Civil Engineering

a First floor plan

Grid with columns A–H (spacing 4 000) and rows 1–3 (spacing 8 000).

- 203 X 203 X 52 kg IC (at A1)
- 457 X 152 X 60 kg IB (beam A1, beam A2)
- column A1, column A2, column A3

Beams along row 1 (between grids): beam 1A, beam 1C, beam 1E, beam 1G — all 533 X 210 X 101 kg IB

Beams along row 2: beam 2A, beam 2C, beam 2E, beam 2G — all 610 X 229 X 113 kg IB

Beams along row 3: beam 3A, beam 3C, beam 3E, beam 3G — all 533 X 210 X 101 kg IB

Vertical beams (between rows 1–2 and 2–3): beam B1, beam C1, beam D1, beam E1, beam F1, beam G1, beam H1; beam B2, beam C2, beam D2, beam E2, beam F2, beam G2, beam H2

b Reference system for floors

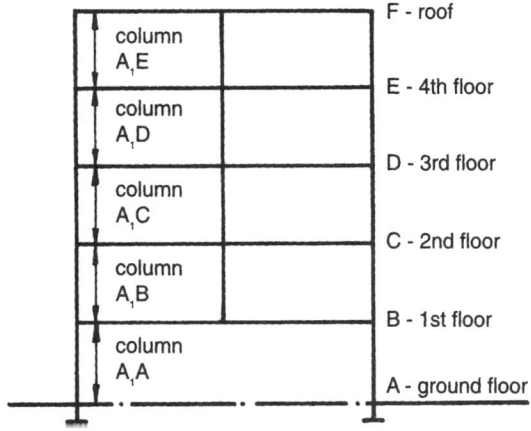

- column A,E
- column A,D
- column A,C
- column A,B
- column A,A

- F - roof
- E - 4th floor
- D - 3rd floor
- C - 2nd floor
- B - 1st floor
- A - ground floor

Fig 6.5 Grid system

position above the adjacent floor level so that the joint, including any splice plates, is well clear of the connections for floor beams.

In Unit 5 you learnt about edge/end distances, gauge distances and bolt pitches. You have to apply this information when determining the size of splice plates.

Fig 6.6 on page 136 illustrates the use of splice plates for columns of different sizes.

6.3 Summary

A column or stanchion is a member whose main function is to carry compressive loading.

Columns are placed vertically in a structure and resist gravity or downward loads. In addition, they are often required to resist bending moments produced by side wind loading or eccentrically applied vertical loading, such as crane loading. There are minor structural steel members in a structure, which resist compressive forces. These members are not necessarily located in the vertical position, and are usually called struts. All columns require a base plate at their lower end to provide the necessary attachment to the concrete foundation, and to transmit the load to the foundation (by means of holding down bolts).

Activity 6.1

(Answers not included)

- A steel column, shown in Fig 6.7 on page 137, consists of a 203 × 203 × 71 kg H-section.
- The column is connected to a 600 mm × 600 mm × 25 mm thick steel base by means of two 150 mm × 90 mm × 12 mm angle iron cleats.
- One cleat is bolted to each flange of the column by means of 4 × 20 mm diameter bolts and each cleat angle is also bolted to the base plate using 4 × 24 mm diameter bolts.
- The base plate rests on 30 mm thick grout and is bolted to the concrete base with 4 × 24 mm diameter holding down bolts.

Remember to apply pitch and edge distances for all the bolts to determine the shape of the cleat!

1. Draw the elevation, looking in the direction of arrow A.
2. Draw a plan elevation of the connection, from (1) above.
3. When you have completed (1) and (2), refer to Unit 5 to check whether the requirements for the end/edge, gauge and pitch distances have been satisfied. Comment on this.

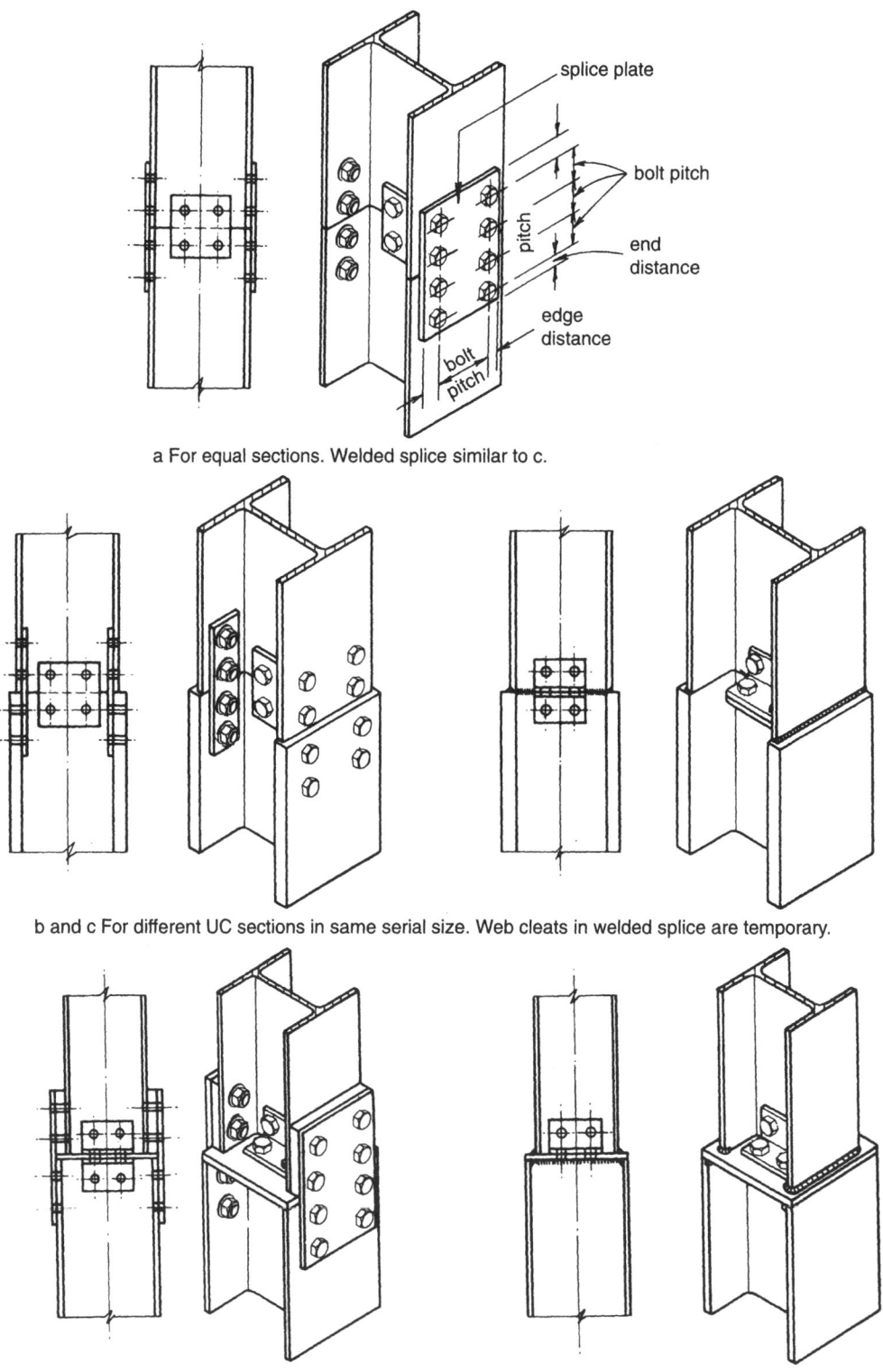

a For equal sections. Welded splice similar to c.

b and c For different UC sections in same serial size. Web cleats in welded splice are temporary.

d and e For different sections. Web cleats in welded splice are temporary.

Fig 6.6 Column splices

Module 2: Unit 6

Draw an isometric view of this column-to-base connection. (This should be a challenge for you!)

Your drawings must be dimensioned to enable trouble-free manufacturing. It is not necessary to draw the concrete base — only show the grout and the top of the concrete base. You should be familiar with drawing concrete bases from Unit 2 in Module 1.

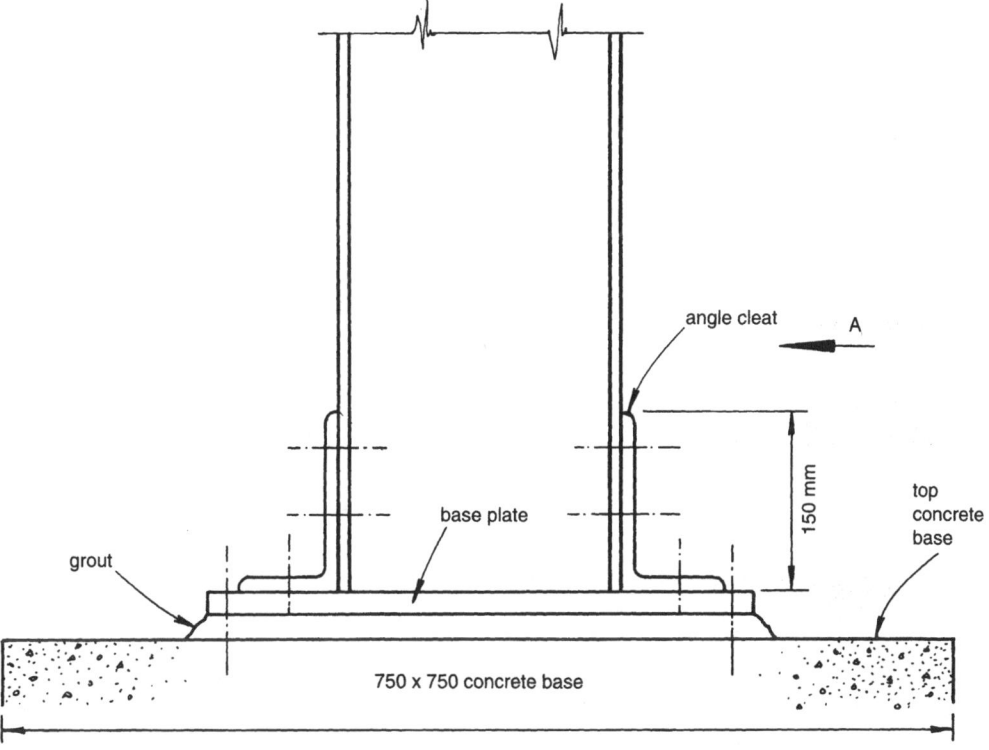

Fig 6.7 A steel column

Self-evaluation

1. Discuss the function of bases.
2. What type of loads is generally transmitted through columns?
3. What is a grid system in detailing columns?

SELF-EVALUATION ANSWERS

1. The function of bases is to transmit vertical loads as well as bending moments into the foundation.
2. Compressive loads are generally transmitted through columns.
3. A grid system is a plan which shows the position and layout of columns, and includes the marking of these columns according to a grid.

Unit 7 Beam-to-column connections

7.1 Beams

In Unit 3 in Module 1, you dealt with reinforced concrete beams, so you should know that beams transfer the load from the floors to the columns. A beam or girder is a member that is subjected to bending. Loading produces a bending moment and shear force. The beam must therefore be of suitable cross-section to resist bending and shear force. The most efficient shape has been found to be the I-section. In most structures the beams are horizontal and resist gravitational loads, that is, loads that act vertically downwards. Examples include main floor beams and secondary floors, which carry the deadweight of the floor plus the live loading of the floor. More examples are the beams in floors, which carry plant equipment and the equipment in industrial buildings. Finally, there are roof rafters that support the purlins and roof cladding of the building (see Unit 9). The connection of a beam to a column serves a threefold purpose, namely:

1. to transmit the vertical forces from beam to column;
2. to hold the beam in place; and
3. to prevent the beam from rotating about its own axis.

If you study Fig 7.1a on page 139, you will notice that the shop bolts and the site bolts that penetrate through the two legs of the angle cleats resist the vertical forces, which 'bridge' between the two members. In Fig 7.1b, a plate welded to the end of the beam replaces the angles.

In Fig 7.1c and Fig 7.1d, the beam rests on a seating cleat, which is fixed to the flanges column, in the workshop. This makes erecting the beam easier. In addition to using the seating cleats it is necessary to use stability cleats, which prevent the beam from rotating upright. Therefore, alternative positions are shown for these in Figs 7.1c and Fig 7.1d.

If you are still a little confused, Fig 7.2 on page 140 will help you to understand these beam-to-column connections.

7.1.1 Notching of beams

Think about the following situation: you have to connect a beam to the web of a column, but the height of the beam is the same as the width of the column. *How do you connect the two?*

You will need to cut away pieces of the flanges of the beam, otherwise those flanges will not fit between the flanges of the column. This process is known as 'notching'. You will see that you are left with dimension 'd' as illustrated in Fig 7.3 on page 141. This 'd' dimension represents the distance between fillets, and it corresponds with h_w in Table 5.1 on pages 102 to 117.

Module 2: Unit 7

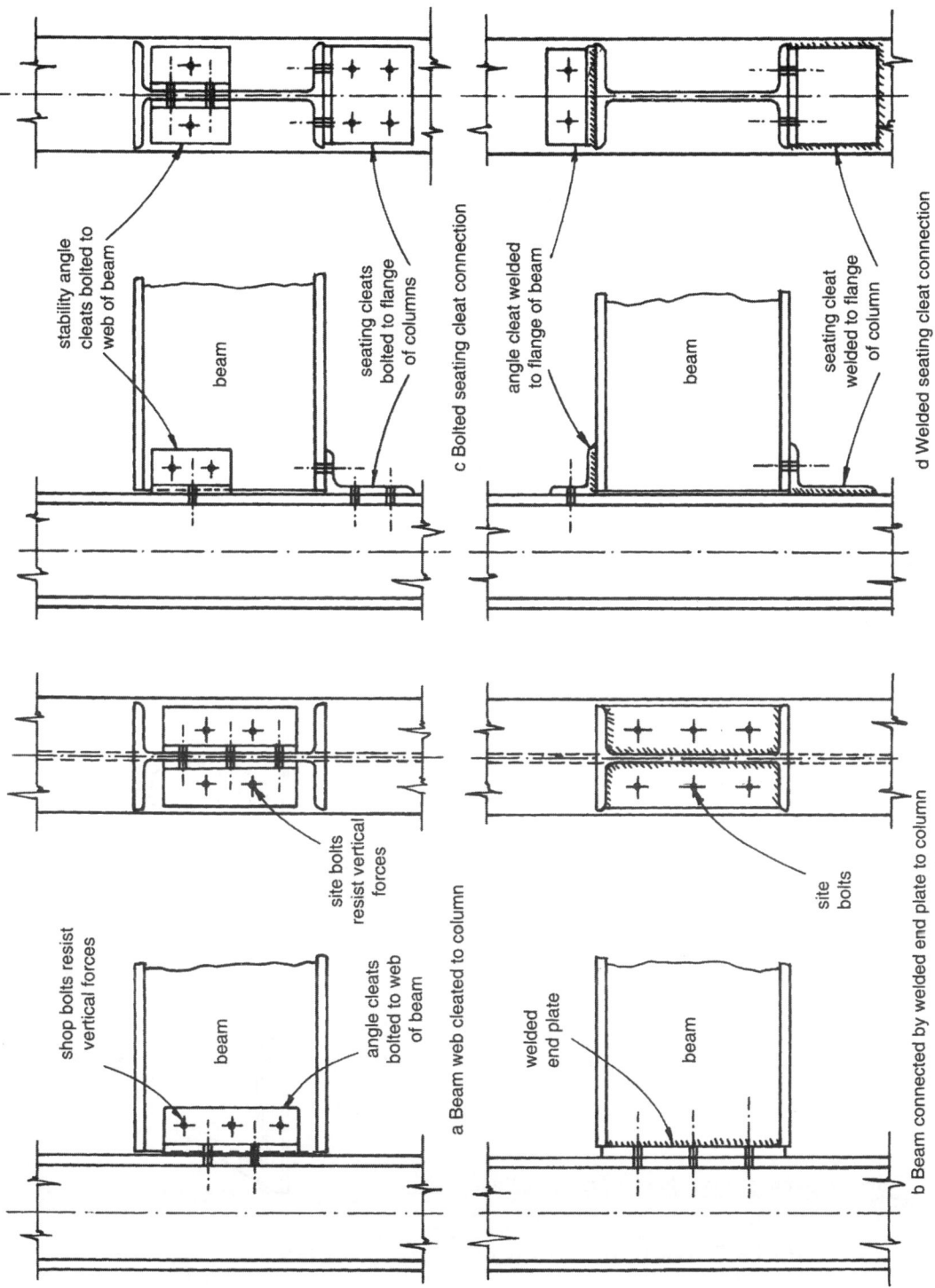

Fig 7.1 Typical beam-to-column connections

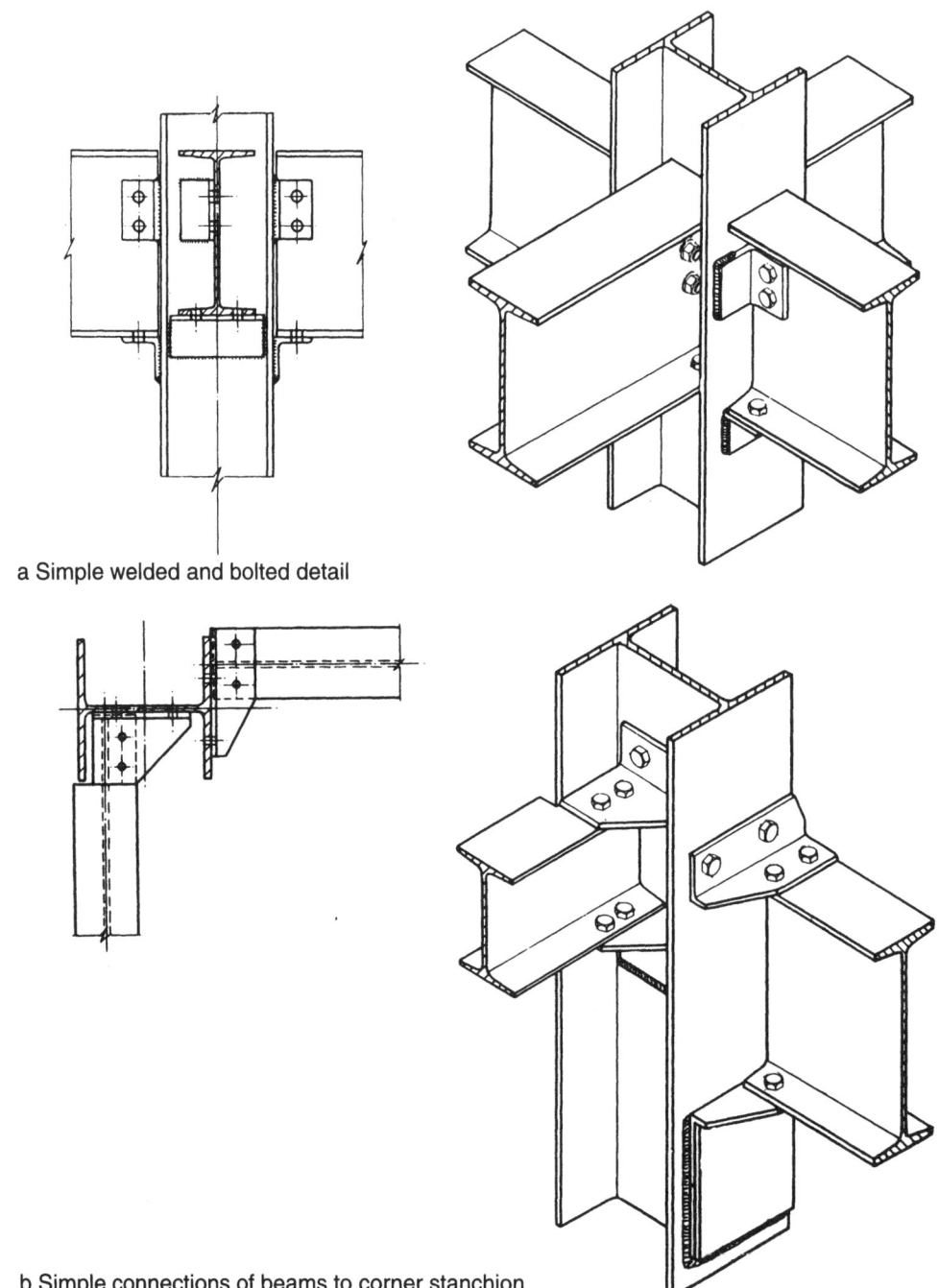

a Simple welded and bolted detail

b Simple connections of beams to corner stanchion

Fig 7.2 Beam-to-column connections

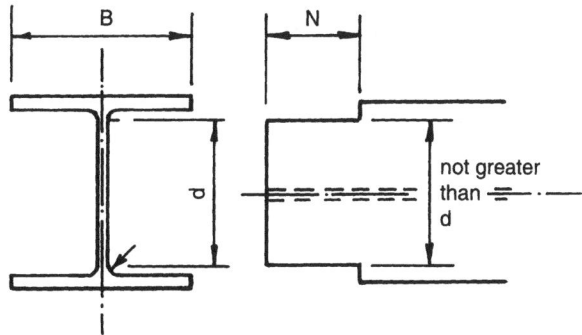

Fig 7.3 Notching of a beam to fit between the flanges of the column

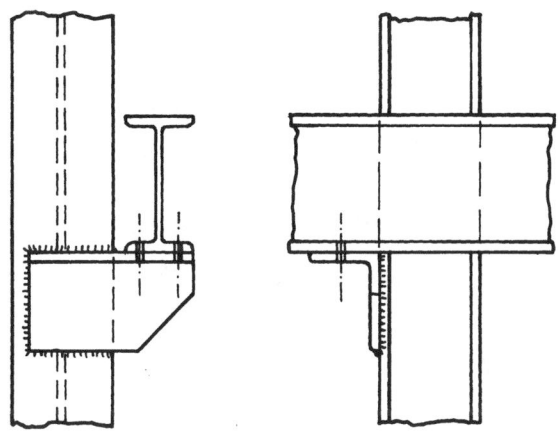

a Eccentric connection using angle bracket

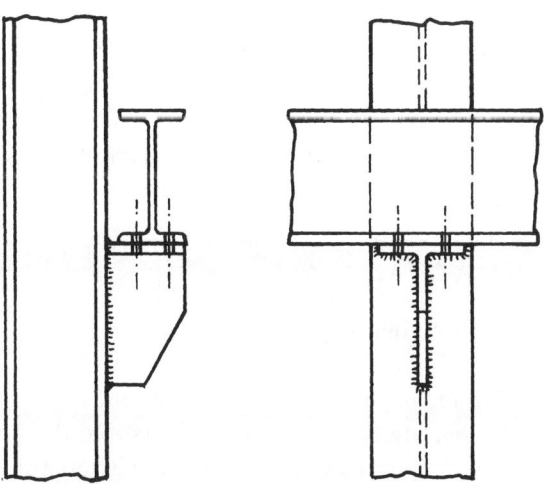

b Eccentric connection using bracket cut out of universal beam

Fig 7.4 Eccentric connections

Perhaps you think that notching would have a weakening effect on the connection; but when notching occurs in this position it does not matter. It would, however, matter if notching occurred at the centre of the **span**, because removing steel at mid-span will weaken the beam.

Remember that notching can only be performed at the connection.

7.1.2 Eccentric connections

As is illustrated in Fig 7.4 on page 141, when a column must support an overhead crane rail, or where beams join the column that does not follow the same centre line, we have a situation where the connection is 'eccentric'. Although the eccentric connections are shown here as welded connections, they could be bolted to achieve the same result. The implications for the structural design of eccentric loading are outside the scope of this book.

7.2 Beam-to-column connections

This beam-to-column section is dealt with by doing Activities 7.1 to 7.3.

7.3 Summary

Beam-to-column connections need to be strong enough to transmit an end moment from a beam to the supporting column. The moment is always accompanied by a vertical shear force and sometimes by an axial force in the beam. The essential requirements for such a connection are adequate strength, sufficient capacity for rotation and ease of fabrication and erection. Bolting the connections adds the advantage of simpler erection. Welded joints are more costly and usually require site welding, but are far stiffer than the bolted connections.

Activity 7.1

(Answers not included)

Study Fig 7.2 on page 140 and identify and label, onto the isometric sketches, the seating cleats, stability cleats and the bolts that resist the vertical forces. Ask your lecturer to indicate where the vertical forces act.

Activity 7.2

(Answers not included)

Fig 7.5 shows an isometric view of a beam-to-column welded connection. Using a scale of 1:5, draw the elevations in first angle orthographic projection, taking the front as seen in the direction of arrow A. Consider all joints to be 8 mm fillet welds.

Does the shape of this angle cleat remind you of the design of the shape of your angle cleat in Activity 6.1?

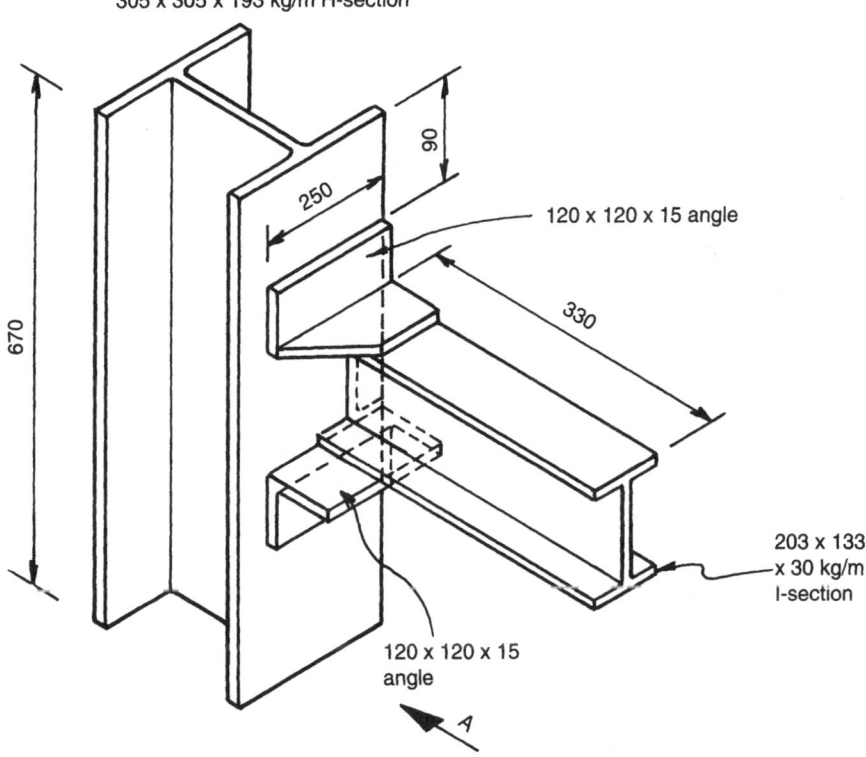

Fig 7.5 Isometric view: beam-to-column welded connection

Drawing for Civil Engineering

Activity 7.3

(Answers not included)

Fig 7.6 on page 145 shows the plan view of a typical beam-to-column connection that occurs at the corner of a structural steel building. The details of the members are given in the diagram. The tops of the beams are at the same level.

The bottom flange cleats (seat angles) are welded to the column and bolted to the beams as follows:
- Beam 'A': 60 mm × 60 mm × 6 mm seat angle with 2 × 16 mm diameter bolts.
- Beam 'B': 120 mm × 120 mm × 8 mm seat angles with 2 × 24 mm diameter bolts.

The web cleat angles are welded to the beams and bolted to the column as follows:
- Beam 'A': 60 mm × 60 mm × 6 mm cleat angles with 5 × 16 mm diameter bolts per cleat.
- Beam 'B': 120 mm × 120 mm × 8 mm cleat angles with 6 × 24 mm diameter bolts per cleat.

Note that each beam will have two cleat angles for the connection to the column. There are no top flange cleats. All the welds are 8 mm continuous welds.
1. Use a scale of 1:5.
2. Draw section A-A and section B-B, showing all dimensions and details, so that the connections could be manufactured in a workshop.
3. Draw an isometric view of this beam-to-column connection. (This should be a challenge for you.)

If you have difficulty in visualising an isometric view, refer back to Fig 7.2a on page 140.

Module 2: Unit 7

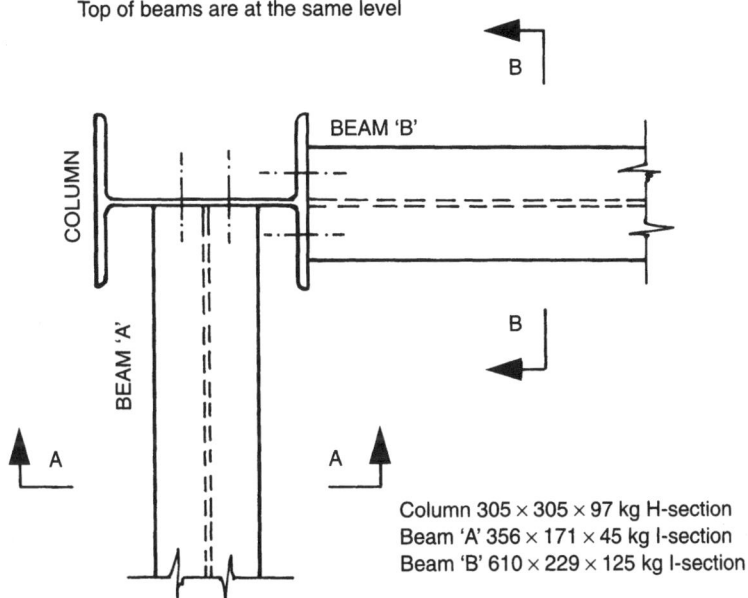

Fig 7.6 Plan at the corner of a building

Self-evaluation

1. Discuss the function of beams.
2. What do you understand by the notching of beams?
3. Describe eccentric connections.

SELF-EVALUATION ANSWERS
1. Beams transfer the load of floors to the columns.
2. Notching involves the removal of pieces of a beam's flange to ensure that it will fit between, for instance, a column's webs.
3. Eccentric connections are beams that join the column other than on the same centre line.

145

Drawing for Civil Engineering

Unit 8 Beam-to-beam connections

8.1 Beams

We dealt with beams in Unit 7, but when we look at beam-to-beam connections, there are a few more details to consider. Some typical beam details are shown in Fig 8.1 on page 147. Let us look at the way we detail and dimension beams in beam-to-beam connections. In Fig 8.2a on page 148 the incoming beam (2) rests on a seating cleat fixed to the main beam (1). This cleat is dimensioned downwards from the top of the main beam (X). Thereafter the holes in the main (Y_1 and incoming beams (Y_2 and Y_3) and the notch (Z) are all dimensioned from the seating surface upwards. These rules also apply to beams resting on seating cleats attached to columns. In Fig 8.2b and Fig 8.2c on page 148 there are no seating cleats, and all the holes (Y_1, Y_2, Y_3, Y_4 and Y_5) and notches (X_1) are dimensioned downwards from the top surfaces of both beams, including the notch in the bottom flange (X_2) in Fig 8.2c.

8.2 Sequential system

Do you remember our discussion about the grid system in columns? We now look at beams in a similar way. We mark beams according to the grid line on which they are positioned. In Fig 6.5a on page 134 of Unit 6, the beams on grid line 1 are marked as 1A, 1C, 1E and 1G, while the beams positioned on grid line A are A1 and A2. The floor number for the beam is indicated by using the floor prefix, thus B-1 A will be on the first floor, C-1 A on the second floor, and so on.

Beams are numbered sequentially — usually starting at the top left-hand corner of the floor plan. These numbers are prefixed by a floor reference, so that beams B1, B2, B3 and so forth are on the first floor; beams C1, C2, C3 and so forth are on the second floor; beams D1, D2, D3 and so forth are on the third floor; and so on. A typical floor plan would look like Fig 6.5a on page 134.

8.3 Splicing beams

Splicing beams is a very similar process to that of splicing columns. Sometimes it is necessary to transport beams in two pieces and to join them together at site, by means of a splice plate. In Fig 8.3 on page 149 we illustrate two ways of doing this. The joint in Fig 8.3a will transmit bending and shearing forces, but the joint in Fig 8.3b only shearing forces.

Module 2: Unit 8

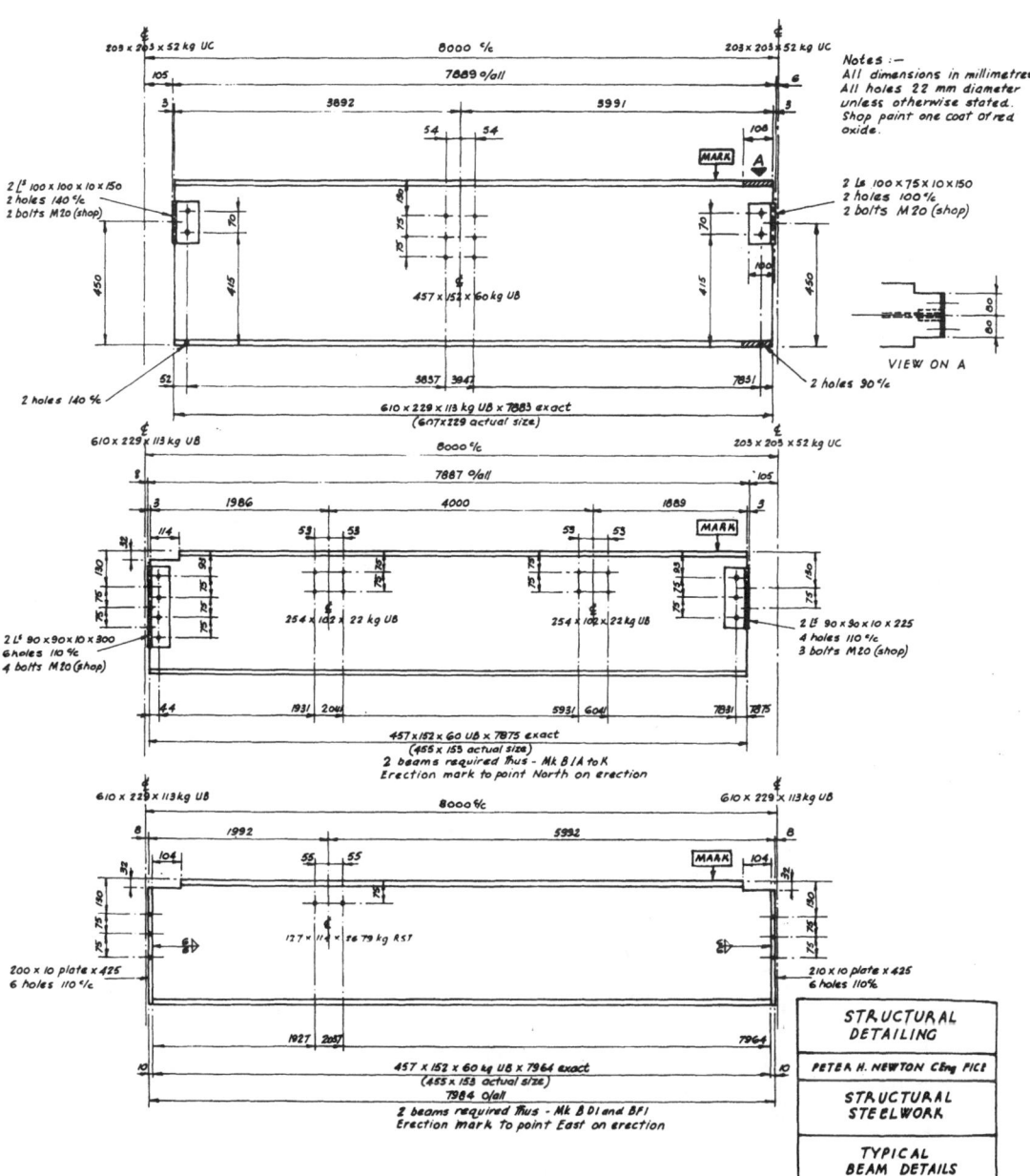

Fig 8.1 Typical beam details

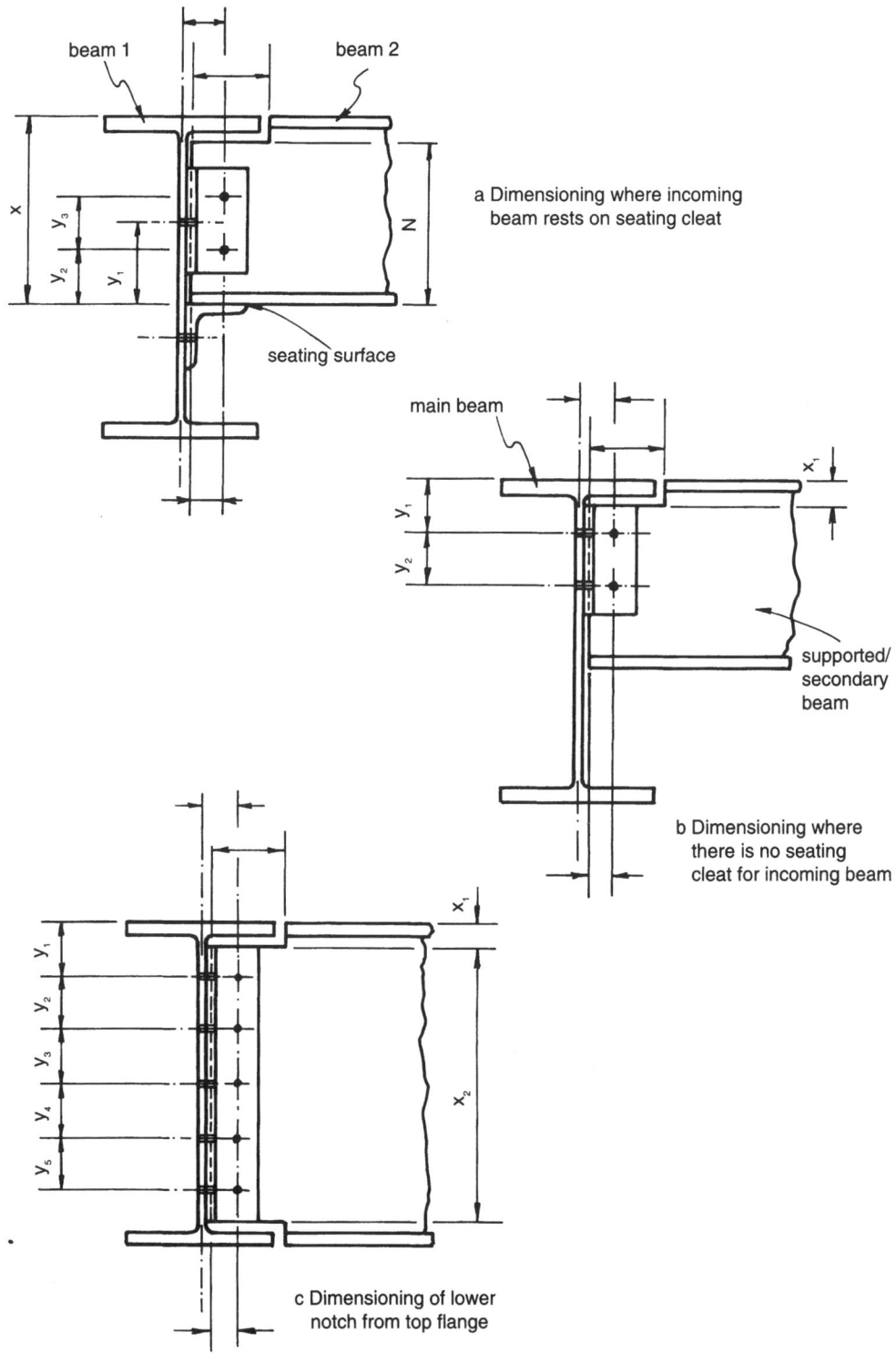

Fig 8.2 Dimensioning of beam-to-beam connections

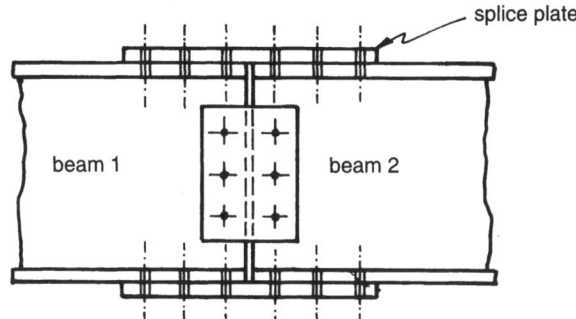
a Joint capable of transmitting bending and shearing forces

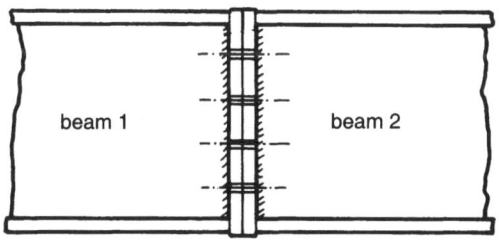
b Joint capable of transmitting shearing forces only

Fig 8.3 Splice joints for beams

8.4 Ways of connecting beams to beams

You will discover that beam-to-beam connections display similarities to beam-to-column connections. In Fig 8.5a on page 151, bolts through angle cleats transmit the vertical forces. In Fig 8.5b, the vertical forces are transmitted by bolts through a flat plate welded to the end of the incoming beam. In Fig 8.5c the incoming beam rests on a seating cleat that is fixed to the supporting beam and is provided with stability cleats. Fig 8.5d suggests one of many ways of dealing with beam-to-beam connections where the tops of the beams are at different levels.

In Fig 8.4 on page 150 you can see examples of web cleats or end plates, which are used to transmit the loads from the secondary beams to the main beam. Seating cleats (brackets) are more economical to use, because they provide a landing bracket (see Fig 8.5c).

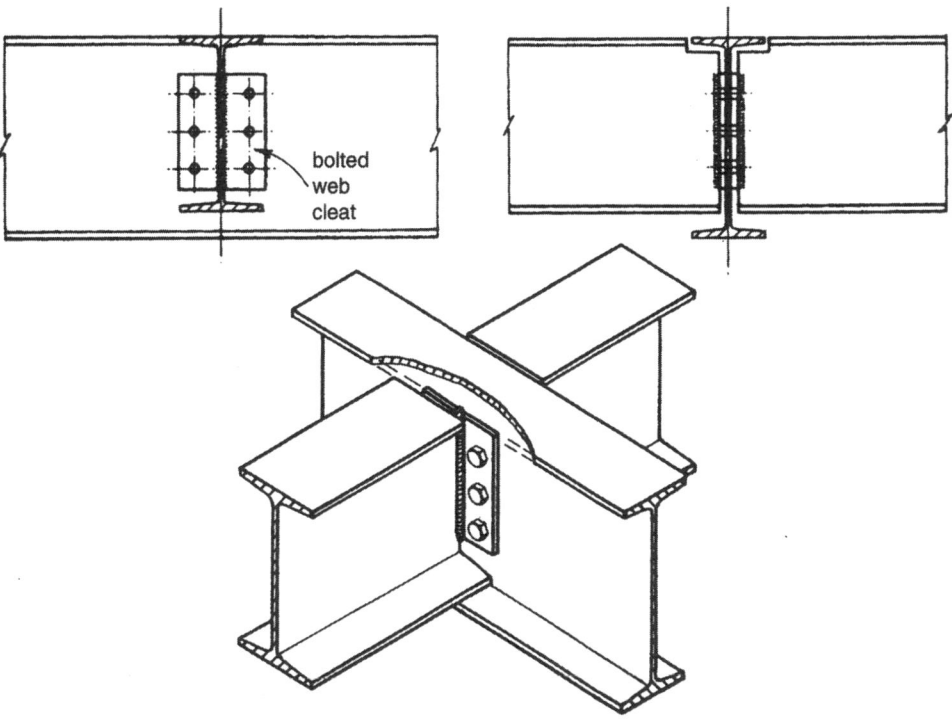

a Secondary beams with welded end plates bolted to main beam

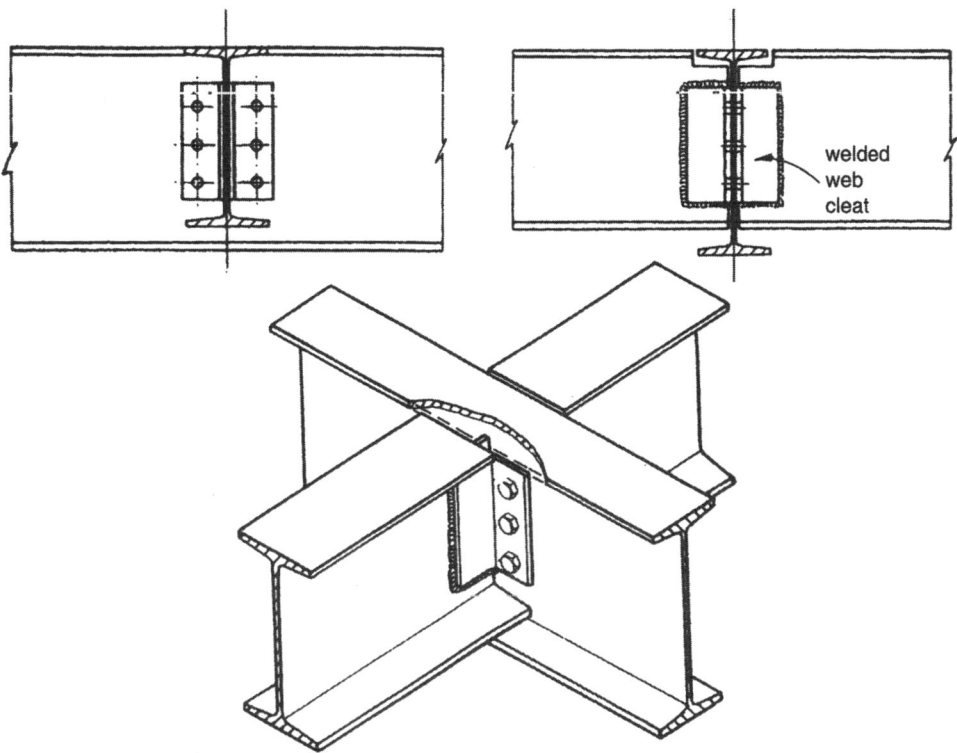

b Secondary beams with welded end cleats bolted to main beam

Fig 8.4 Beam-to-beam connections

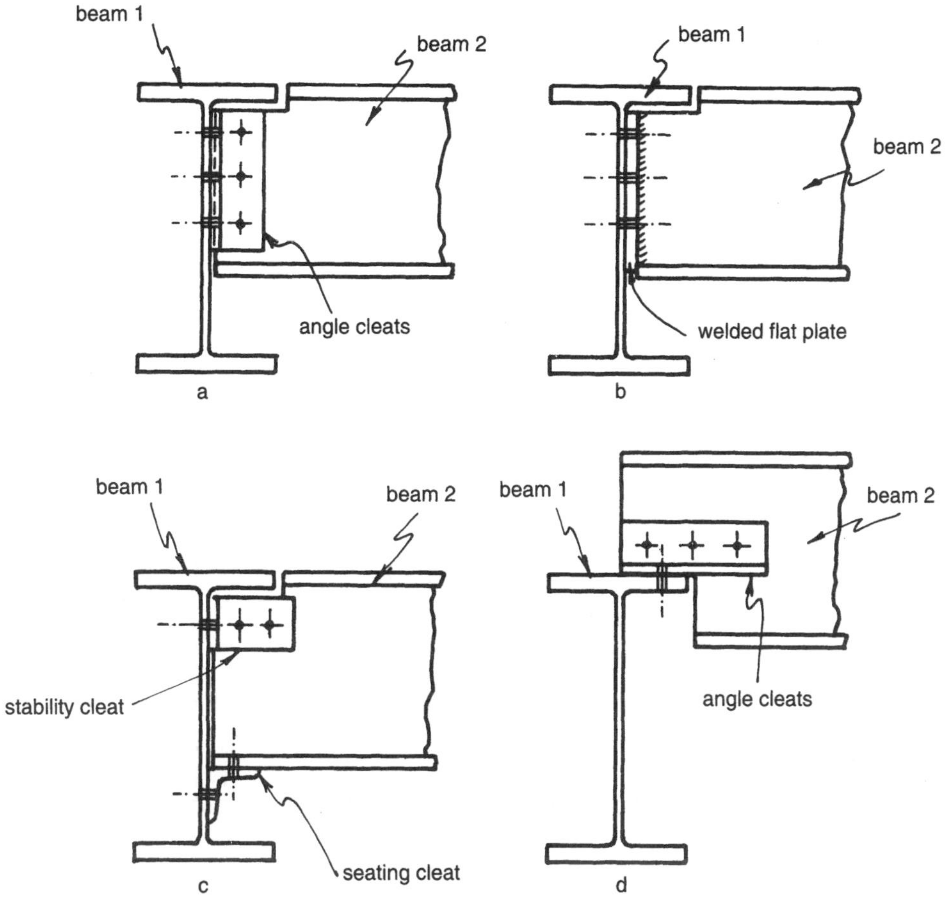

Fig 8.5 Ways of connecting beams to beams

8.5 Summary

A beam is usually supported at each end by relying on either a column or a main beam, which requires typical beam end connections. In beam-to-beam connections the connecting elements can be bolted angle cleats, welded end plates or plates welded to the supporting members. The top flange of the supported beam is cut back or notched in order to clear the top flange of the supporting beam.

Activity 8.1

(Answers not included)

The part floor plan and the elevation of the steel for a multi-storey building are shown in Fig 8.6 on page 152. Make detail drawings for beams B1, B2 and B3. Use a scale of 1:25. All particulars are shown on the drawing.

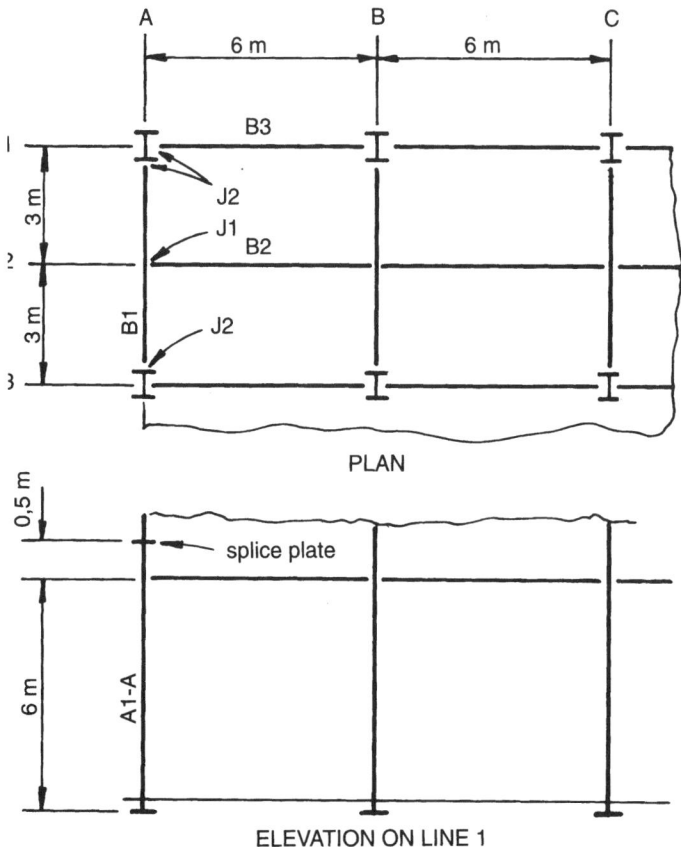

Beams
B1: 406 × 140 × 46 kg/m I-Beam
B2, B3: 356 × 171 × 45 kg/m I-Beam

Joints
J1: 6 No. 20 mm dia. black bolts
J2: 6 No. 20 mm dia. black bolts

Stanchion
A1: 203 × 203 × 46 kg/m H-Column

Base slab
550 × 550 × 38

Fig 8.6 Part floor plan

Self-evaluation

1. What do you understand by a sequential system?
2. Why is it necessary to use splices in beams?

SELF-EVALUATION ANSWERS
1. A sequential system is similar to the grid system for columns. Beams are numbered sequentially.
2. Sometimes it is necessary to transport long beams in two pieces, and then join them again on site.

Unit 9 Roof structures

9.1 Introduction to roofs

All houses have roofs that protect people from the sun, the rain and the cold. *How many of you have seen the construction of a roof?* In *Theory of Structures* we discussed this as frames. There we calculated the nature of forces and the forces in each member. If you need to remind yourself, go back to the relevant section in *Theory of Structures*. You should know that the roof construction consists of members constructed in such a way that they are able to carry the roof covering material as well as to resist the forces of nature, such as rain, wind, snow or hail.

 Do you remember this discussion?

We are going to discuss the steel frames that are commonly used in roof structures, namely roof trusses, lattice girders and portal frames.

9.2 Roof trusses

We can describe a roof truss as a plane frame that consists of (a) sloping rafters, which meet at the ridge; (b) a main tie connecting the feet of the rafters; and (c) internal bracing members, as shown in Fig 9.1. Roof trusses and purlins support the roof covering. You might be wondering, 'What are purlins?' Purlins are secondary members connected longitudinally across the rafters. The roof covering is attached to the purlins. See Fig 9.1.

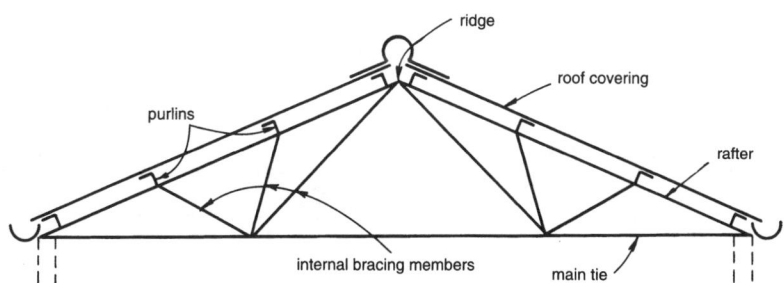

Fig 9.1 Roof trusses

Let us look at the arrangement of the internal framing of a roof truss. Note that this arrangement depends on the span of the roof truss. Rafters are normally divided into equal panel lengths. Ideally, the purlins should be supported at the panel points, so that the internal bracing members are only subjected to axial forces. This is unfortunately not always practicable. In fact, you should know that purlin spacings vary with the type of roof covering that is being used. It is not unusual for purlins to be supported between panel points (so that the rafter members have to be designed to withstand local bending action, in addition to axial forces). Fig 9.2 helps to clarify these concepts.

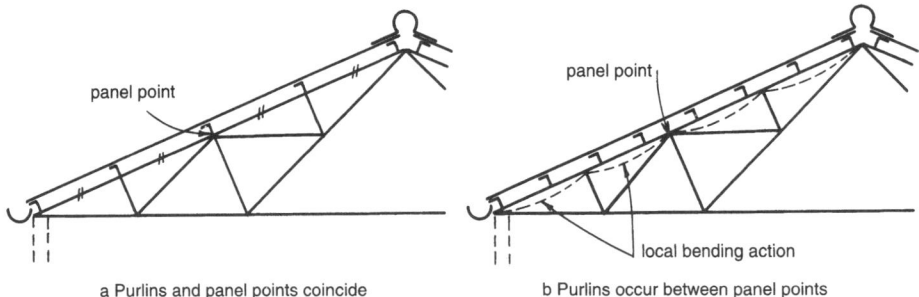

a Purlins and panel points coincide

b Purlins occur between panel points

Fig 9.2 Purlins and panel points

Maximum purlin spacings associated with different types of roof covering (cladding) can vary from about 900 mm to over 3,5 m. Let us look at the internal bracing members. The long members are generally in tension (ties), and the shorter members in compression (struts). Fig 9.3 shows some typical roof trusses and the spans for which they are most suitable.

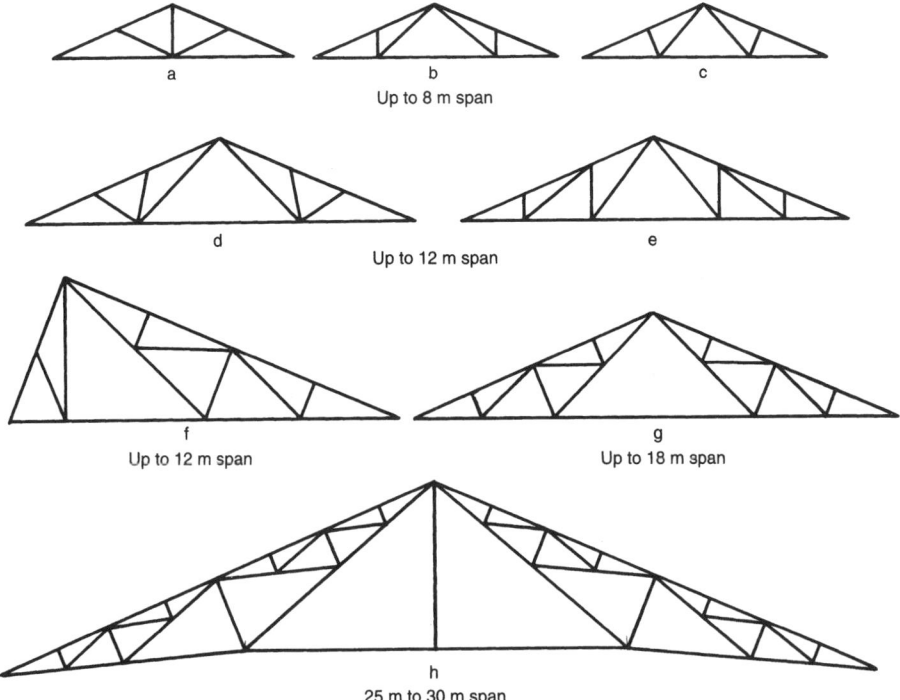

Fig 9.3 Typical roof trusses

What about spans of more than 30 m? Conventional trusses with very long spans and rafters at constant slope are not often acceptable. This is because of their excessive height and the additional difficulties in solving drainage and heating problems. To overcome these problems, the mansard-type truss illustrated in Fig 9.4 may be used.

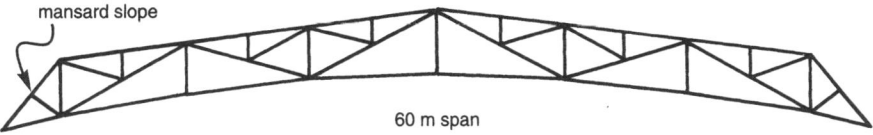

Fig 9.4 Mansard-type truss

Using steep-sloping (or even vertical) members at the end of the truss as well as a shallower slope over the rest of the span reduces the ridge height. The long span trusses, shown in Fig 9.5, are in the order of a 70 m span and a rise of almost 20 m; they are generally spaced every 15,5 m.

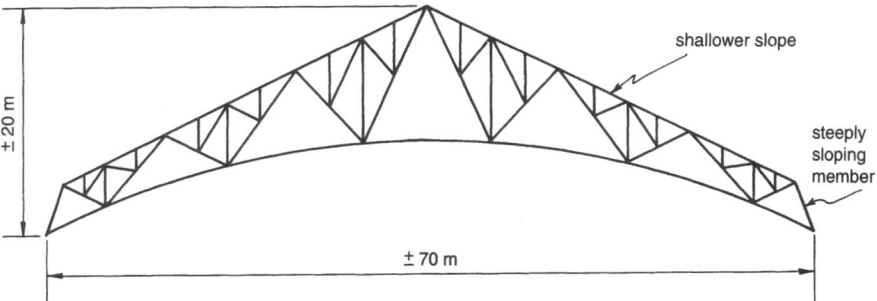

Fig 9.5 Long span trusses

Originally, rivets were used as a means of joining the members of a truss. However, with the development of welding, the fabrication method illustrated in Fig 9.6 on page 156 superseded riveting. Welding has two advantages:
1. It replaces gusset plates (see 'bolted connections' for an explanation of gusset plates).
2. It is neater and less complicated.

Another way of connecting internal bracing members is by bolting them together. This is more economical than welding. A typical bolted roof truss is shown in Fig 9.7 on page 157.

Let us compare the differences at the nodes when considering welded (Fig 9.8) and bolted (Fig 9.9) roof trusses. Look at Fig 9.8 (Detail 2) on page 158. Notice that the heel of the internal bracing is positioned at the top. When you look at Fig 9.9 (Detail 2) on page 159, you will notice that the heel is positioned at the bottom. There is no particular design rule to determine this.

Drawing for Civil Engineering

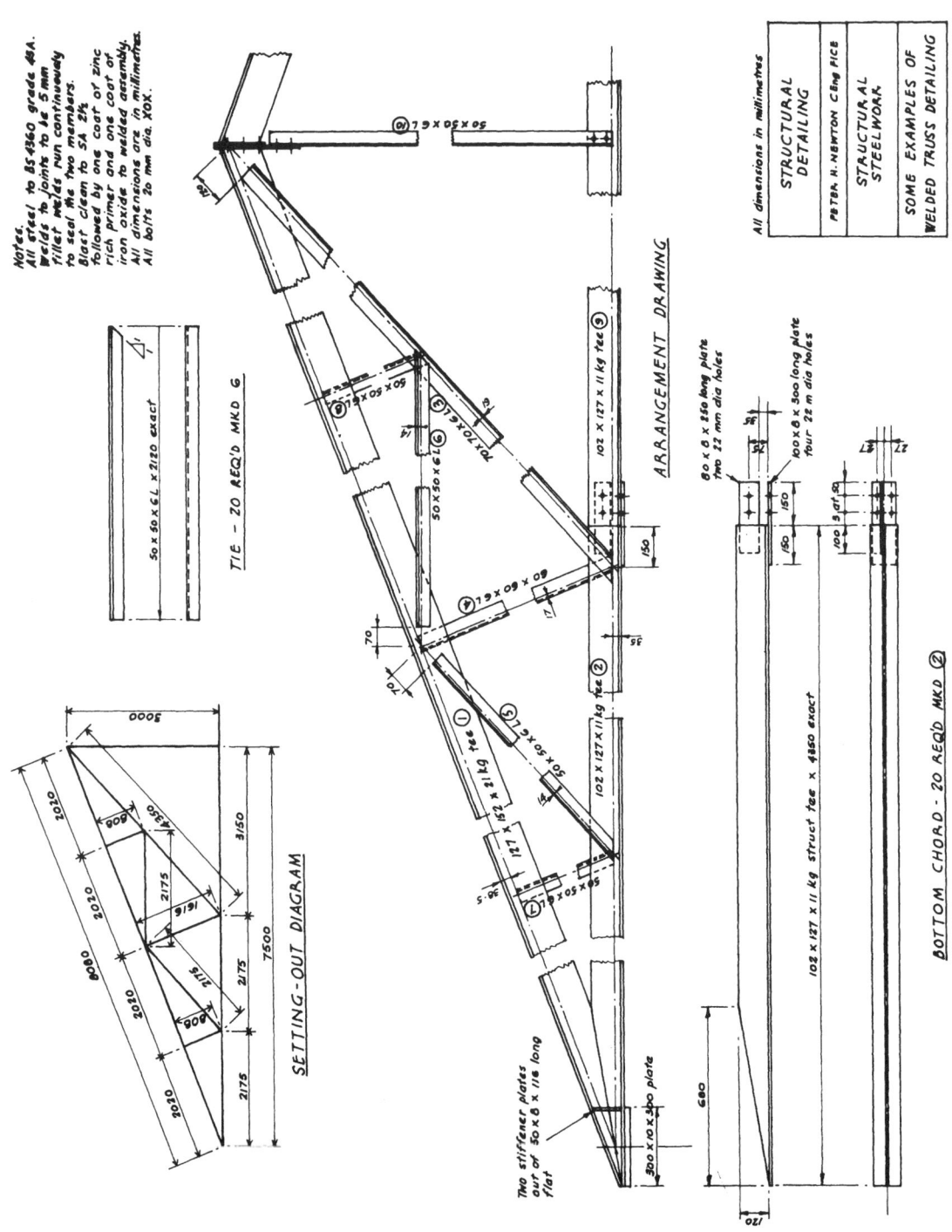

Fig 9.6 Welded truss detailing

Module 2: Unit 9

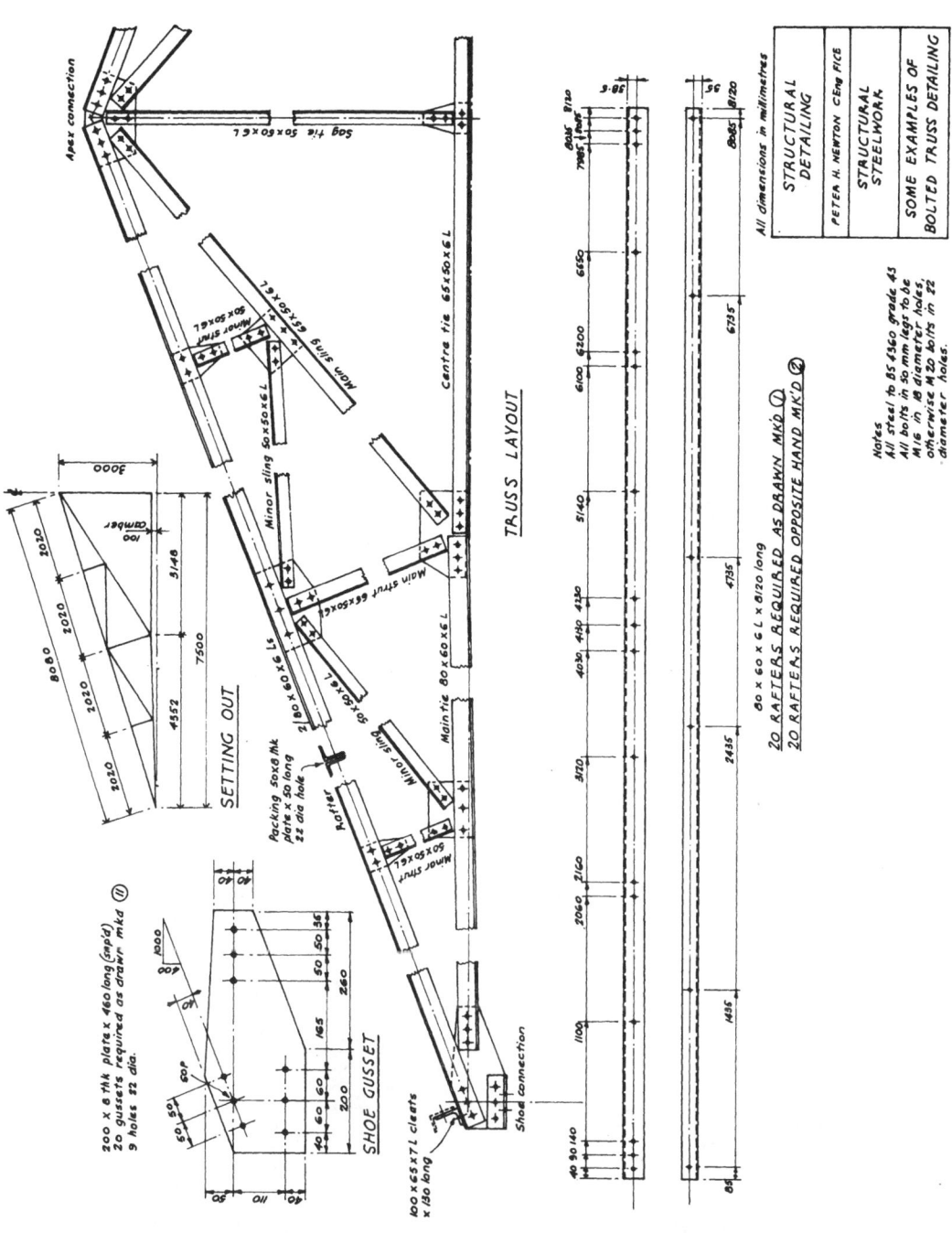

Fig 9.7 Typical bolted truss detailing

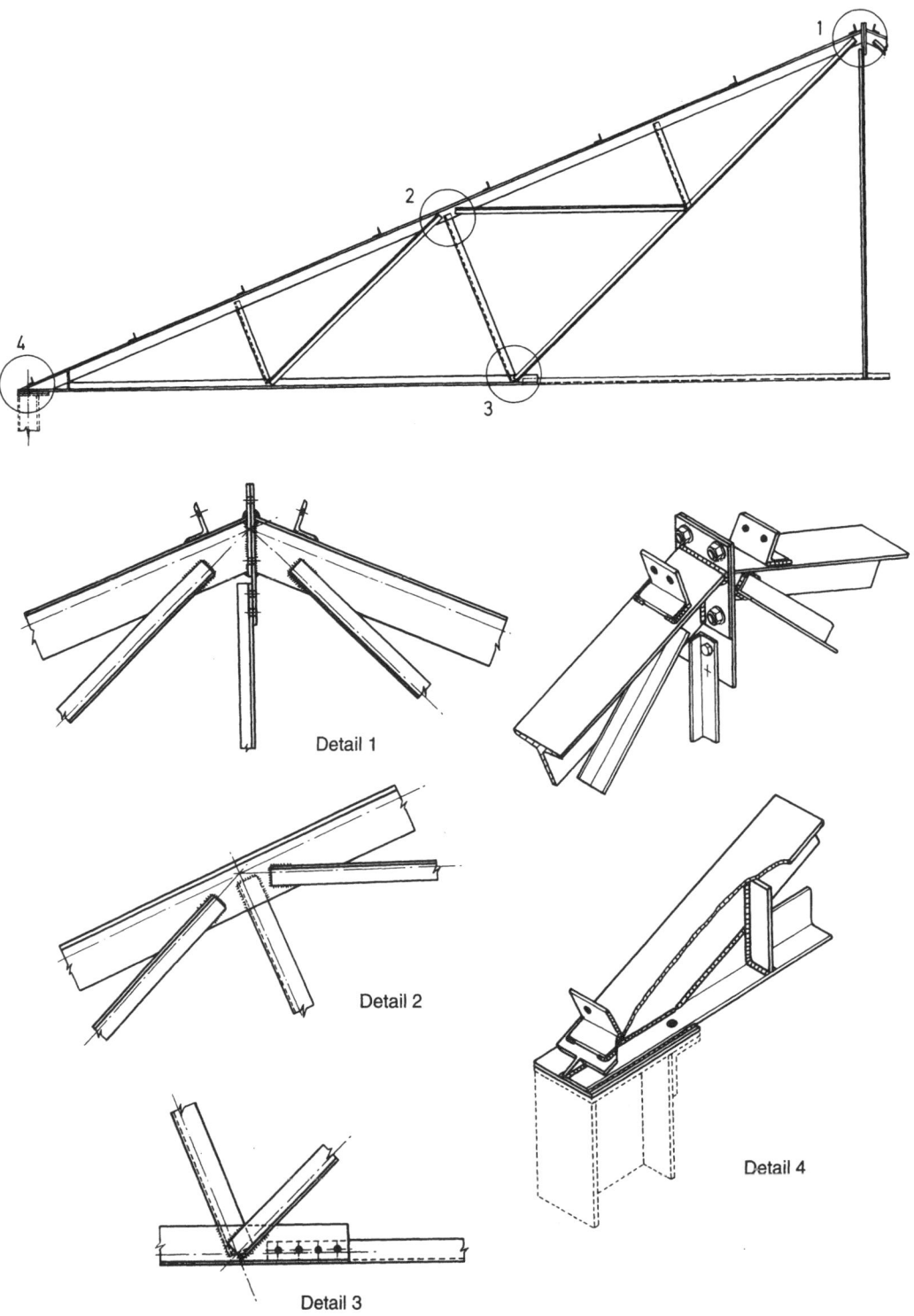

Fig 9.8 Welded roof truss

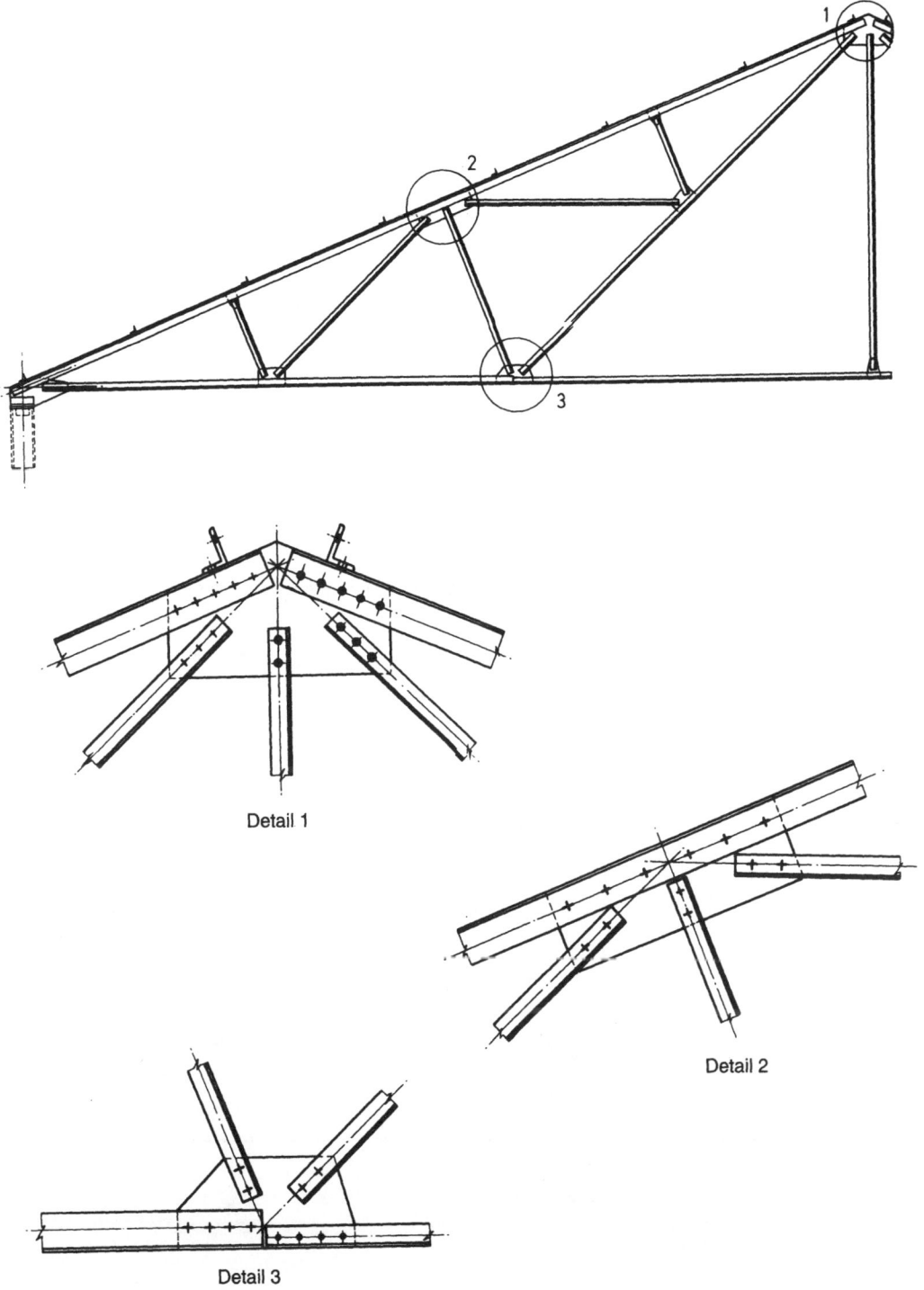

Fig 9.9 Bolted roof truss

9.3 Lattice girders

How does the construction of lattice girders differ from the construction of roof trusses? Lattice girders are frames of open web construction, and usually have parallel chords (and parallel booms) when used for roofs and with internal web bracing members. They are extremely useful in long span construction where the depth-to-span ratio is small, generally from about 1/10 to 1/14. This gives lattice girders a distinct advantage over roof trusses. There are two main types of lattice girders, namely the N-type (as can be seen from the internal member arrangement) shown in Fig 9.10a, and the Warren type shown in Fig 9.10b.

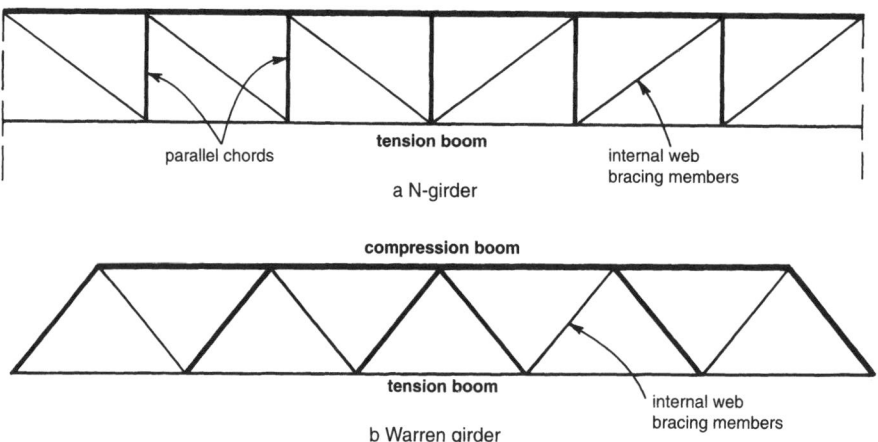

Fig 9.10 Lattice girders (Note: The thick lines indicate struts.)

Where it is essential to omit diagonal members, the 'rigid joints' girder, shown in Fig 9.11, can be used where all the internal members are vertical and the joints are made rigid. What is meant by 'rigid joints'?

 To find the answer, review your *Theory of Structure* notes.

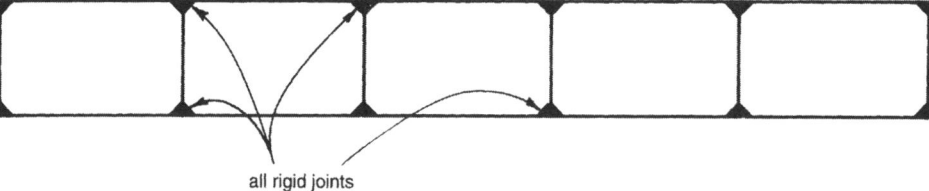

Fig 9.11 'Rigid joints' girder

Remember, as mentioned earlier, that when we determine the forces in the members of frames, we assume that the external loads are applied at the node points or joints, and that all members meeting at a point are represented by lines that intersect at that point. It therefore makes sense that the members should be set out so that the lines of their centres of gravity meet at the node points. Note that in the case of bolted connections, centre-of-gravity lines would coincide with the backmark positions for angles. This common point, where the member axes meet at a node or panel point, is called a setting-out point (SOP).

Fig 9.12 shows that for bolted connections, standard bolt pitches and end distances are used where possible. Also, a 10 mm clearance is provided at member ends. Fig 9.13 on page 162 shows the extent of the welding. Fig 9.14 and Fig 9.15 on pages 163 and 164, respectively, also explain this concept for both bolted and welded nodes.

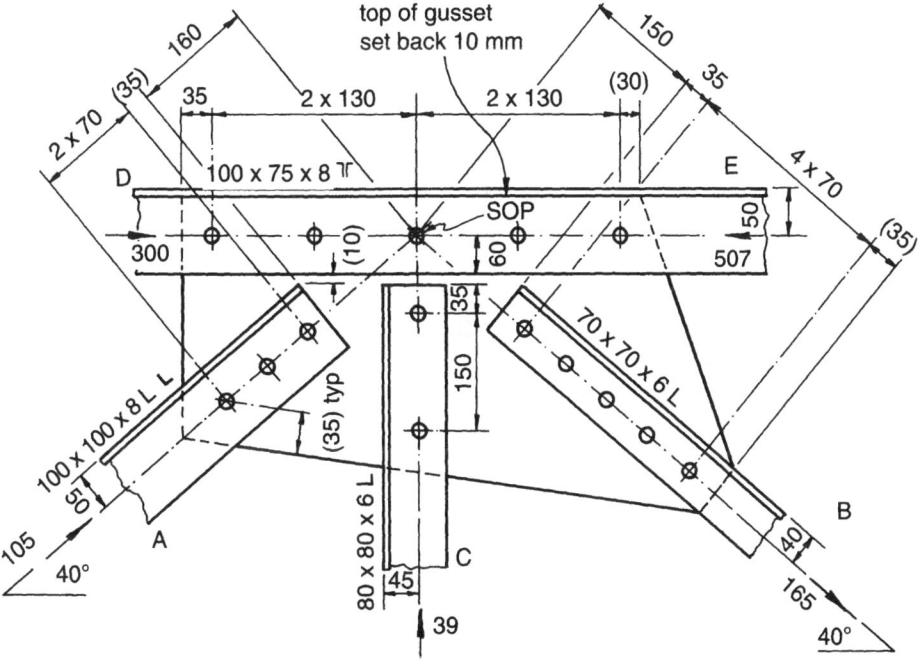

Fig 9.12 Layout of nodes — bolted construction

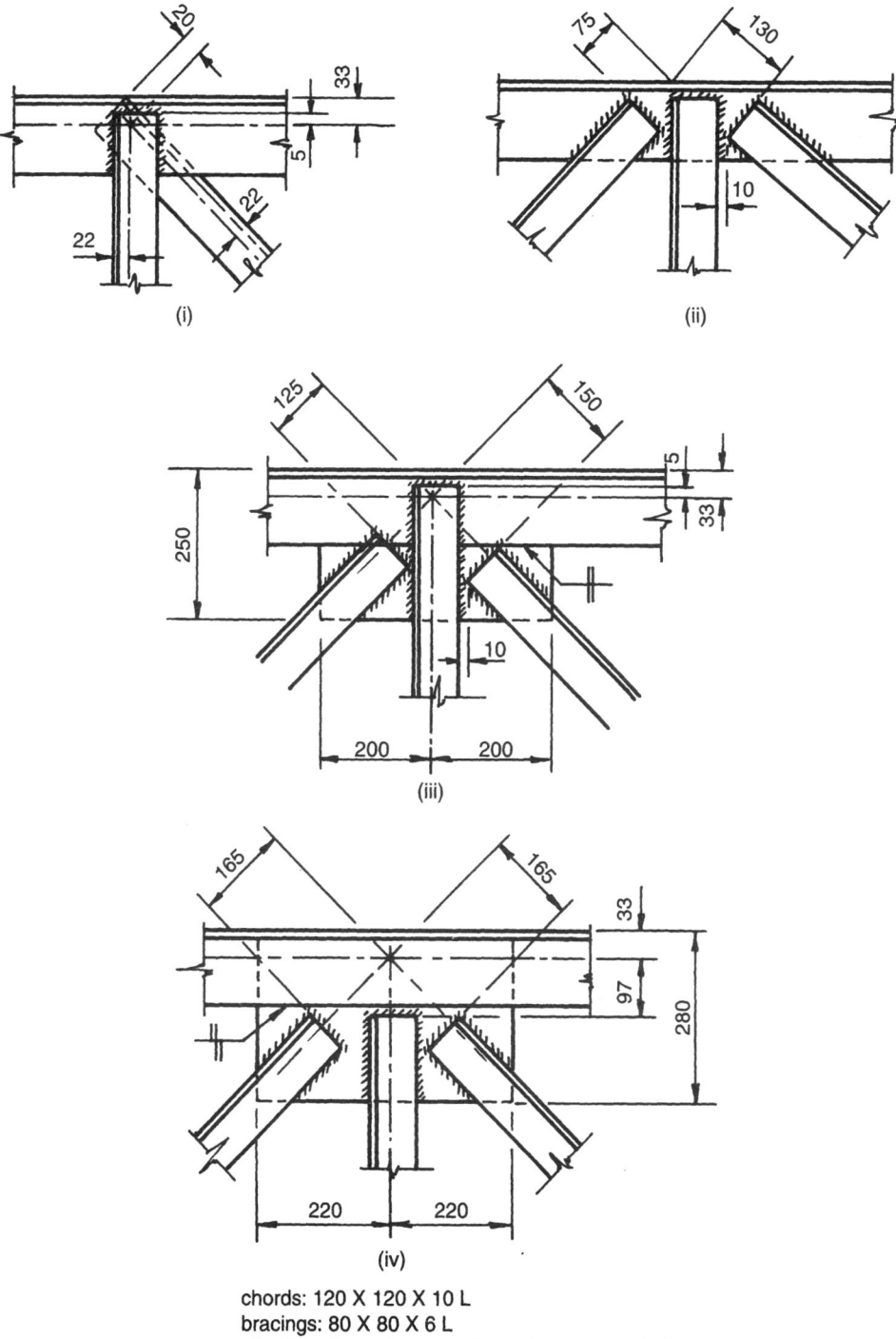

chords: 120 X 120 X 10 L
bracings: 80 X 80 X 6 L
all set backs and gusset dimensions are scaled

Fig 9.13 Layout of nodes — welded construction

Module 2: Unit 9

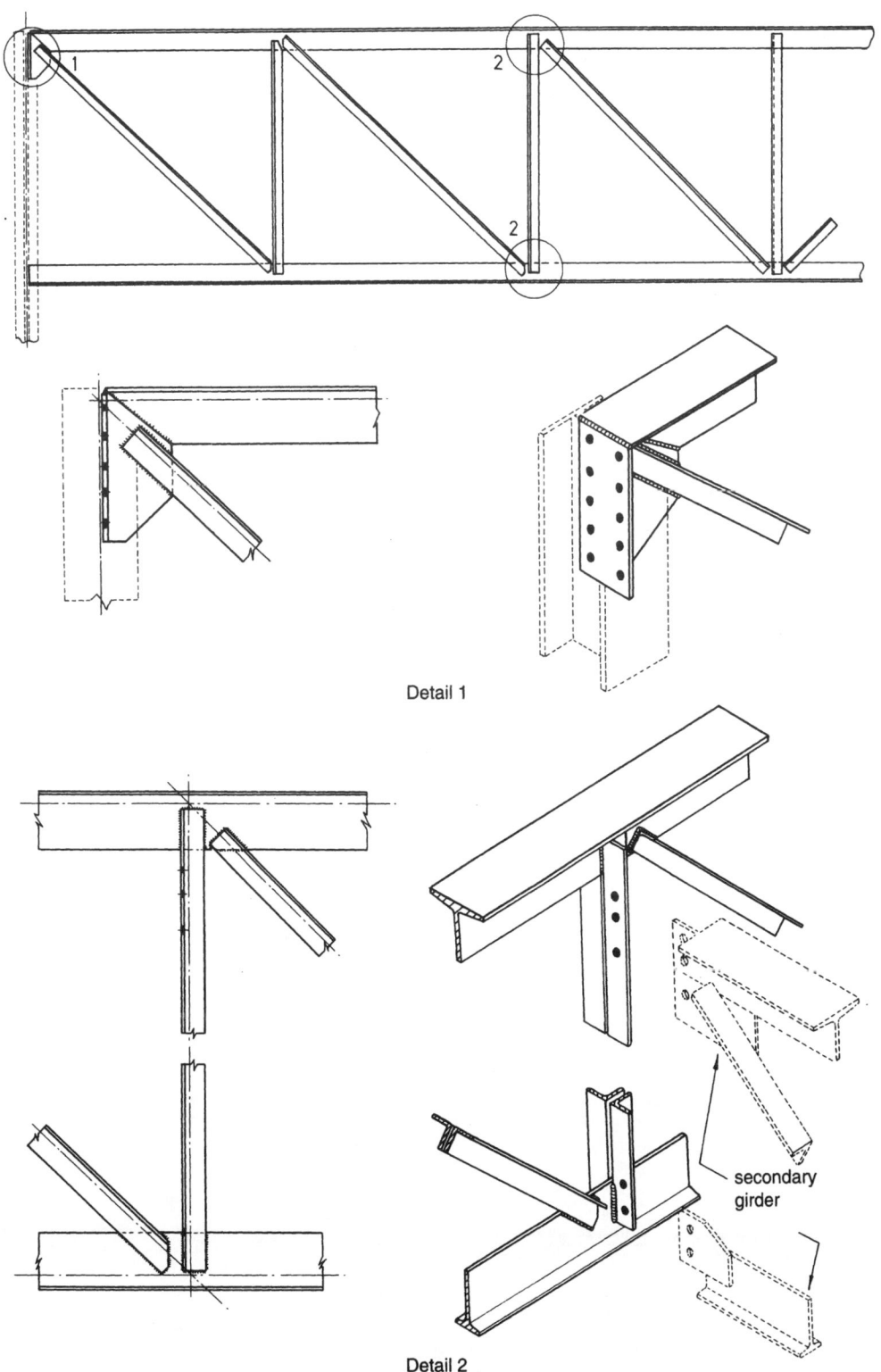

Detail 1

secondary girder

Detail 2

Fig 9.14 Welded lattice girder

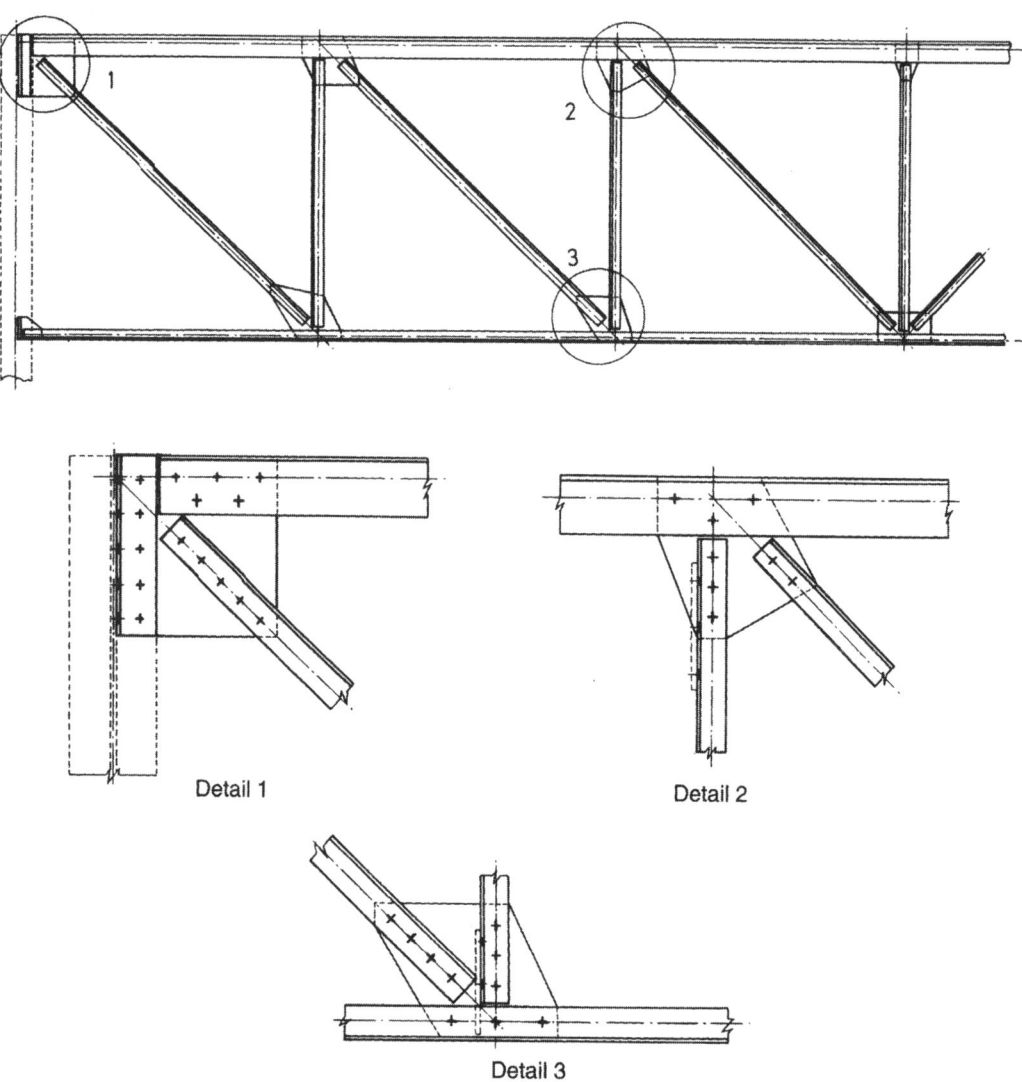

Fig 9.15 Bolted lattice girder

Module 2: Unit 9

9.4 Portal frames

The third type of roof structure is the portal frame. A portal frame is a type of arch construction in which the roof member, whether a horizontal beam or pitched rafter, is joined rigidly at the eaves to the stanchion to form a continuous plane frame (see Fig 9.16a). The pitched roof portal frame has the advantage of providing clear working space from floor to rafter level, which is unobstructed by ties or bracing members. Portal frames can be of solid or open web construction, as illustrated in Fig 9.16b.

The portal frame that is shown in Fig 9.17 on page 166 is one that is suitable for spans of up to about 18 m and height from ground level to eaves of about 5 m that are arranged at centres up to 4,5 m.

At site, before it is erected, the frame is assembled on the ground, using high-strength friction grip bolts. (Refer to your *S1* notes on rivets and bolts again.) Let us include Fig 9.18a on page 167 and Fig 9.18b on page 168 to illustrate the detailing of different components of a portal frame.

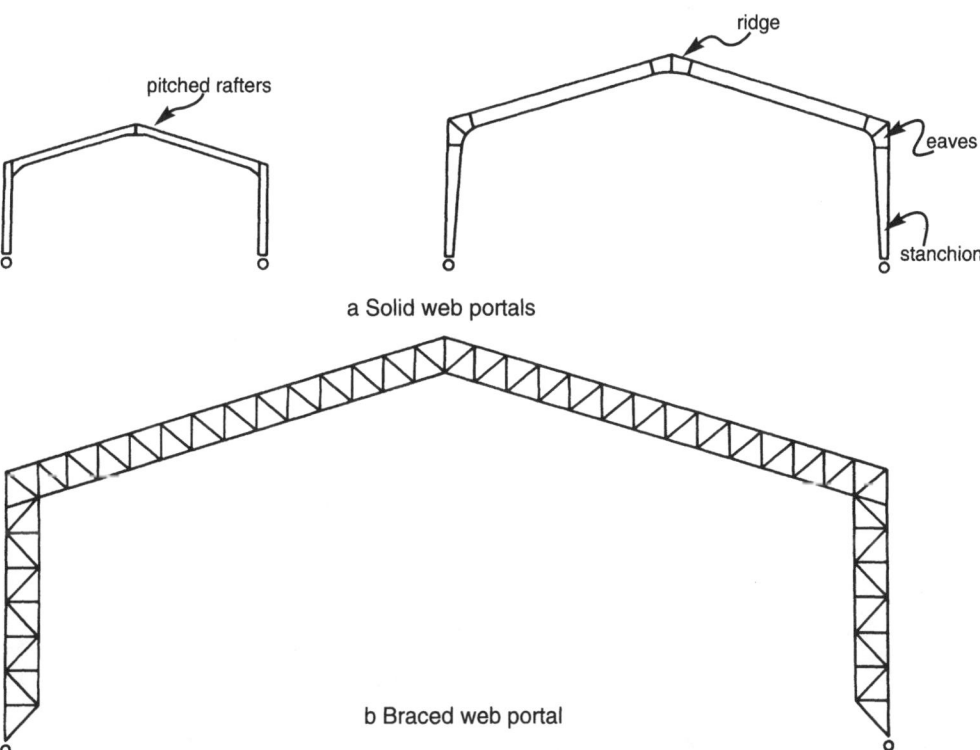

Fig 9.16 Web portals

165

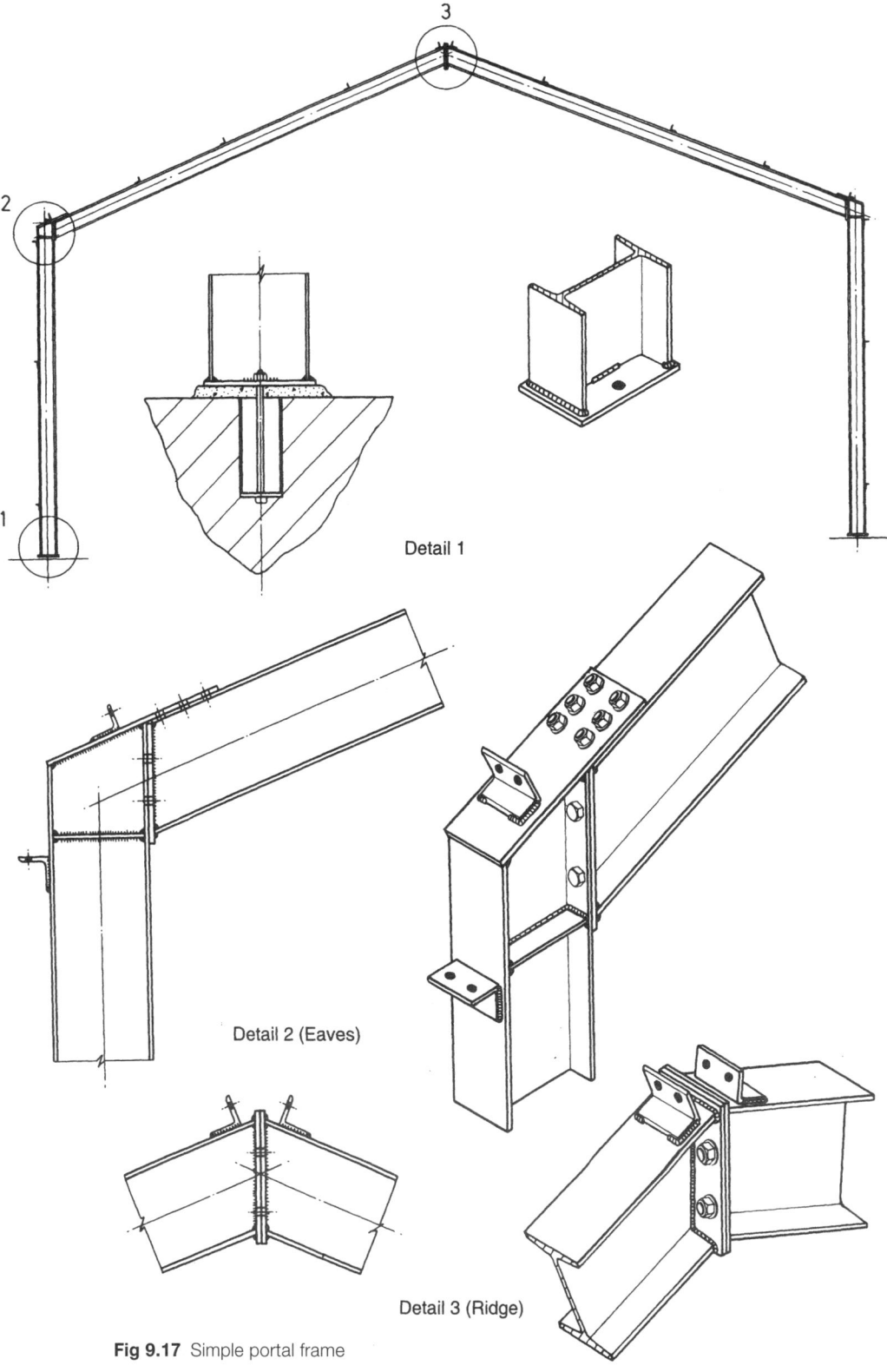

Fig 9.17 Simple portal frame

Module 2: Unit 9

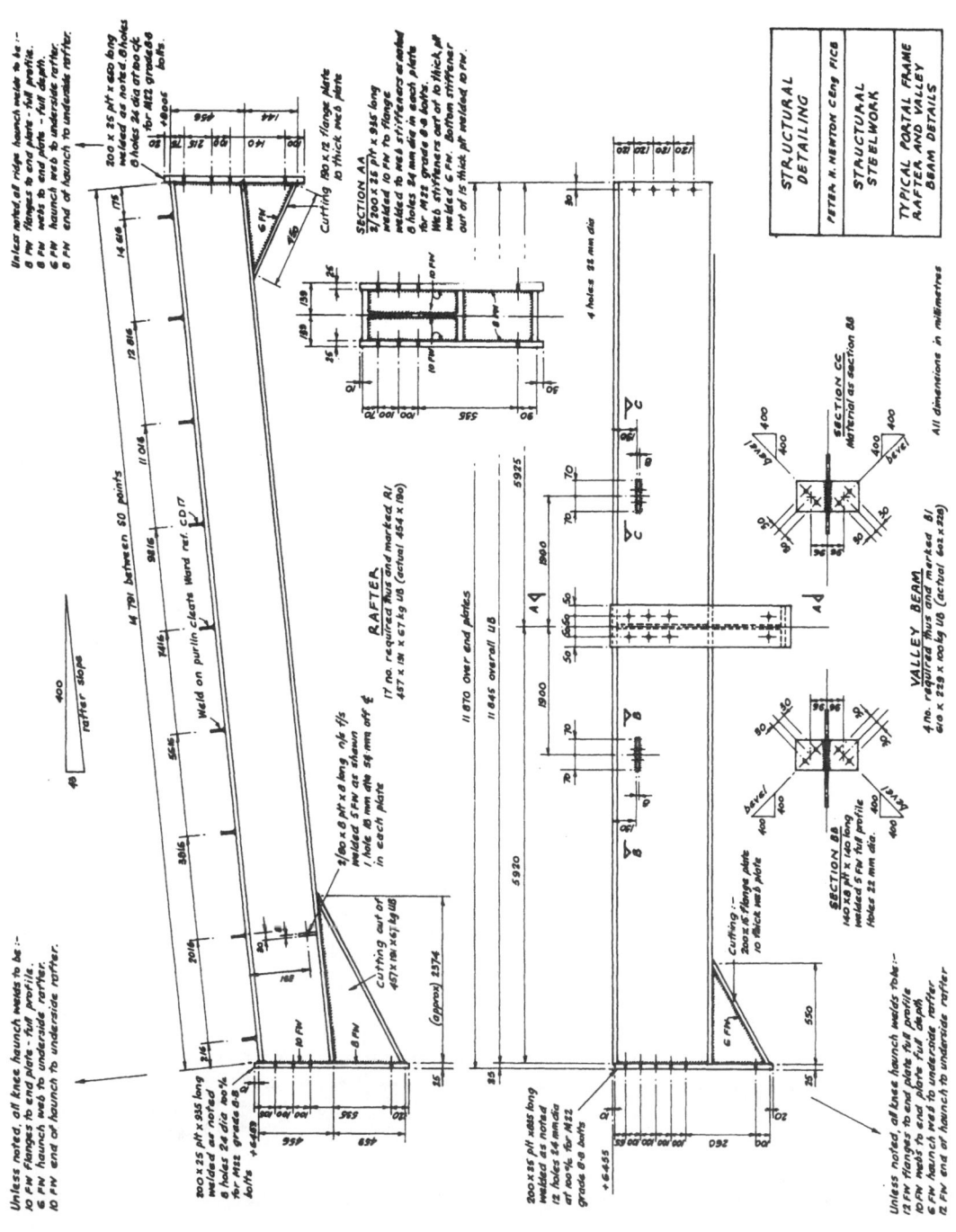

Fig 9.18a Typical portal frame: rafter and valley beam details

Drawing for Civil Engineering

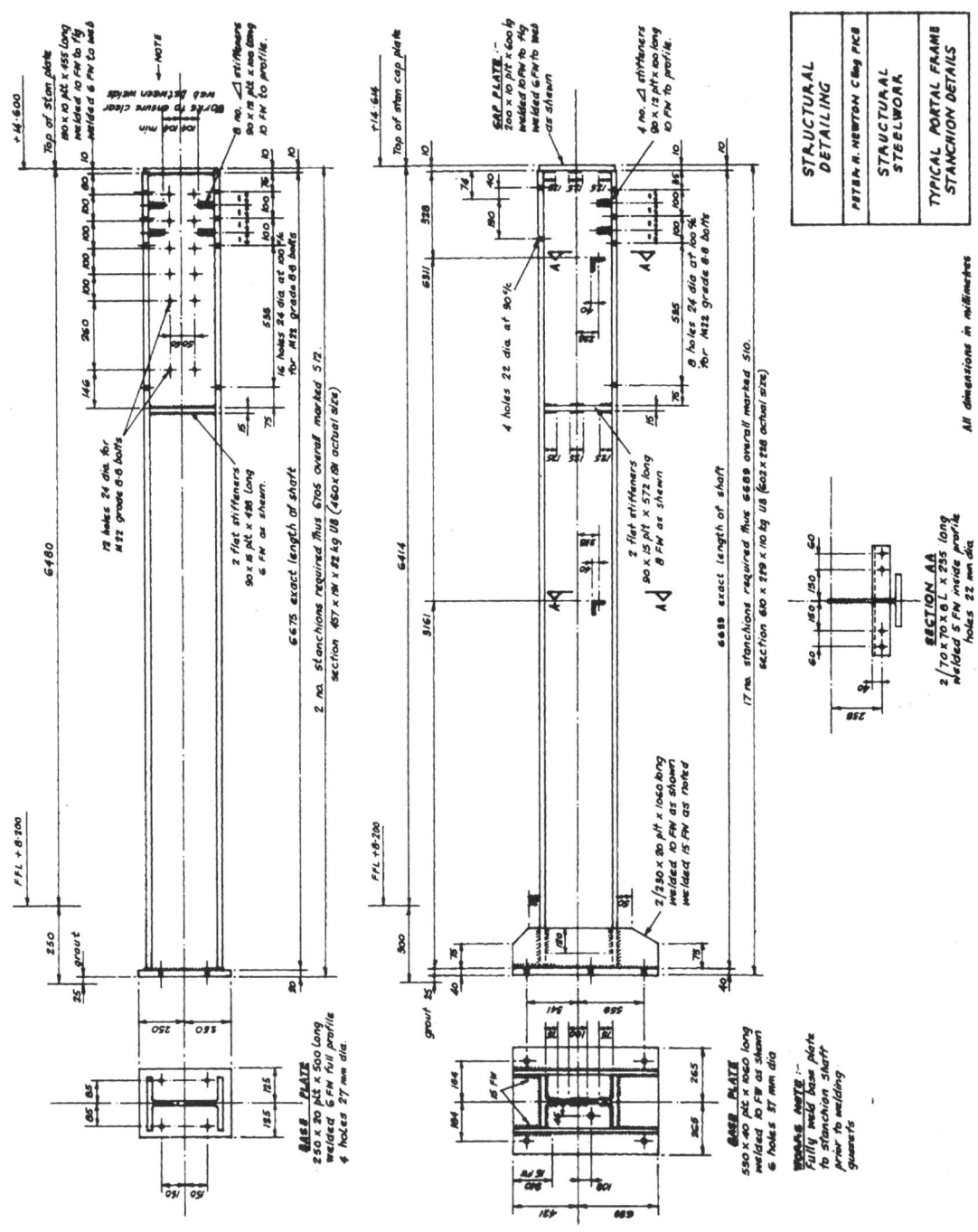

Fig 9.18b Typical portal frame: stanchion details

9.5 Roof systems

Up to this point, we have only considered one plane frame of each type of roof structure, which is only one component of a roof system. However, there are a number of methods to build a roof system.

9.5.1 A typical truss and purlin system

This is shown in Fig 9.19.

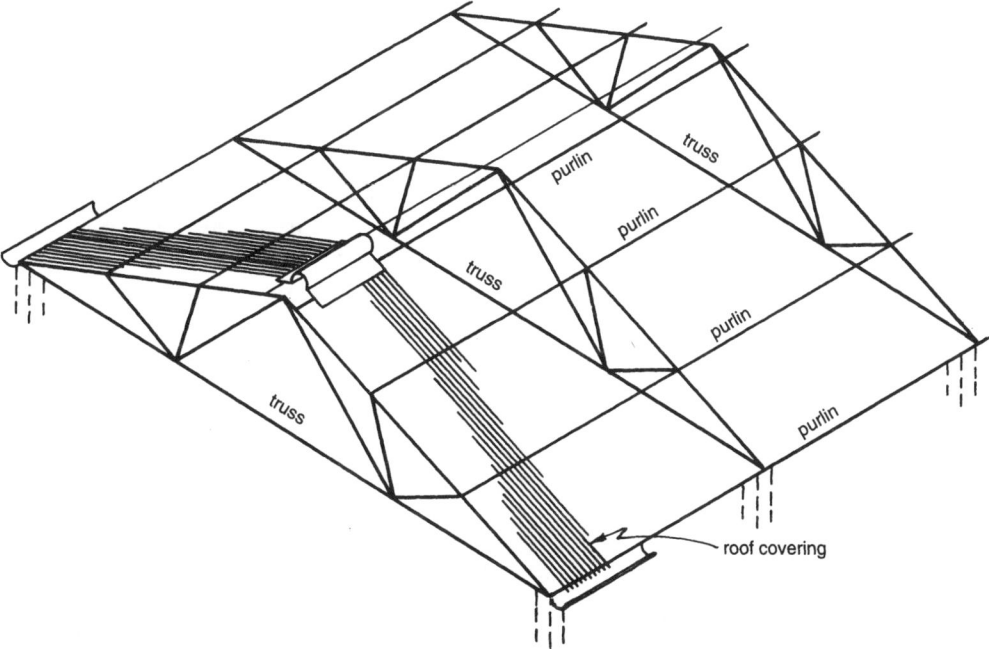

Fig 9.19 Typical truss and purlin construction

9.5.2 Lattice girders

These can be used as shown in Fig 9.20.

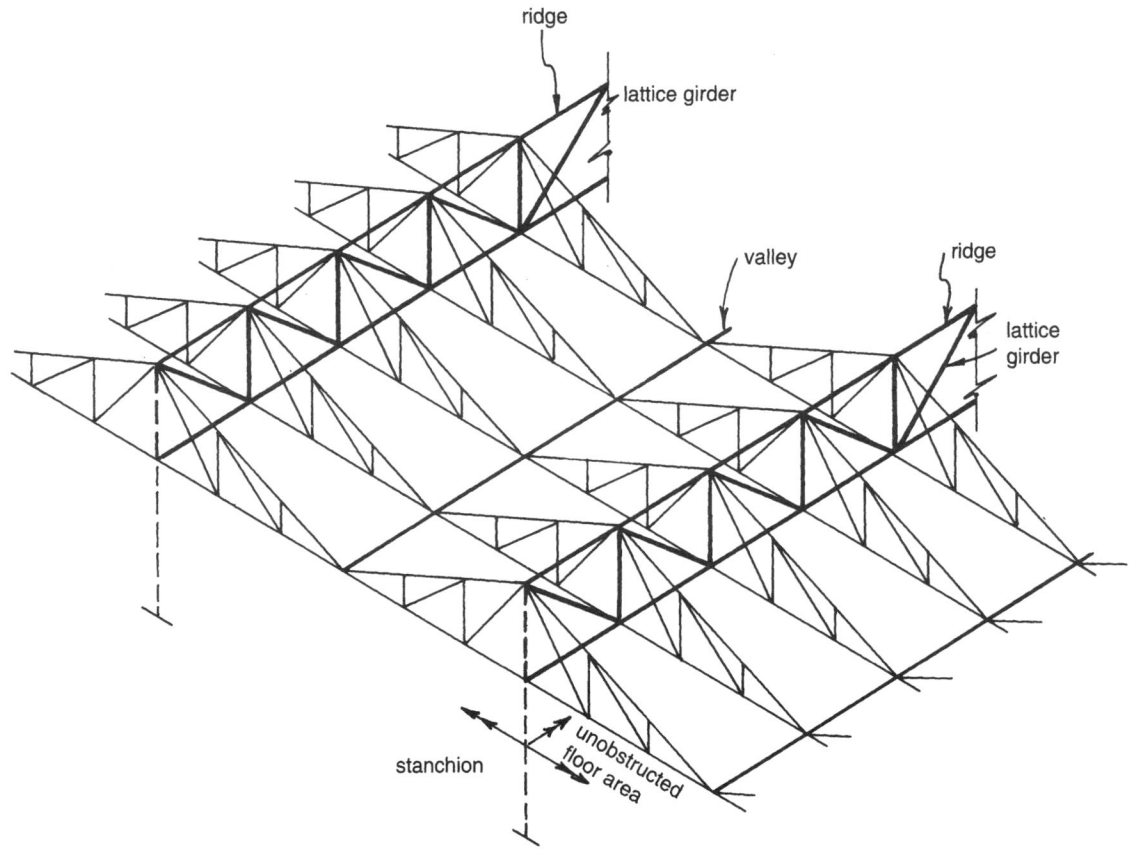

Fig 9.20 Lattice girders

9.5.3 Portal frames

These are shown in Fig 9.21.

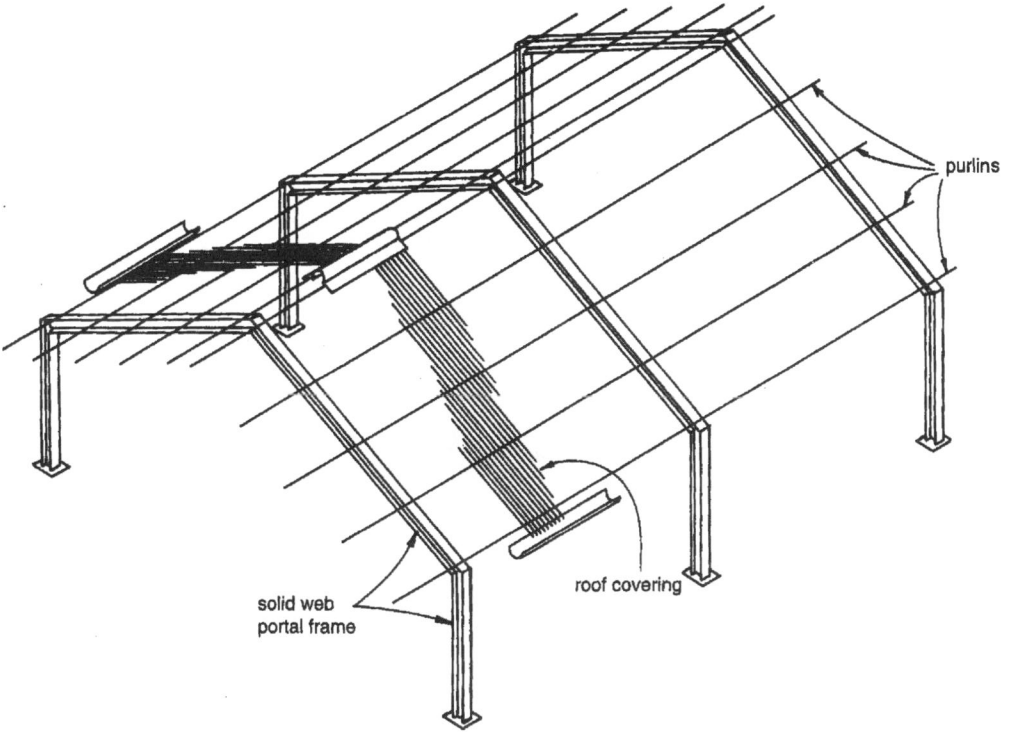

Fig 9.21 Single-bay portal frame

9.6 Summary

Trusses and lattice girders are used in the roof construction of buildings of medium to large span. They are able to support heavy loading on greater spans than I-section beams or rafters. Lattice girders may be used to support floors where wide, uninterrupted space is required below. Lattice girders are also used in footbridges, walkways and the like. Trusses and lattice girders may be welded or bolted construction.

The welded portal frame is one of the most popular forms of construction for small-span industrial buildings. Portal frames are very efficient in material use and are therefore competitive in price. However, portal frames are labour intensive and need to be made accurately. The columns and rafters are made from different sized I-beams that have welded haunches at their ends, which enable the members to be site-bolted to each other to make up the complete portal frame.

Activity 9.1

(Answers not included)

Refer to Fig 9.22 on page 173 and draw the following details to a scale of 1:5:
1. Detail 1 — a welded connection;
2. Detail 2 — a bolted connection; and
3. Detail 3 — a bolted connection.

You can assume the following:
- 2 × 16 mm bolts per connection as a minimum;
- The beam and rafter sections are 80 mm × 80 mm × 8 mm angle;
- Other members: 60 mm × 60 mm × 8 mm angle;
- Backmarks are to be the theoretical value; and
- Column to be shown as follows: 203 mm × 203 mm × 86 kg.

H-section
1. Use 10 mm gusset plates.
2. Use 6 mm fillet welds.

Module 2: Unit 9

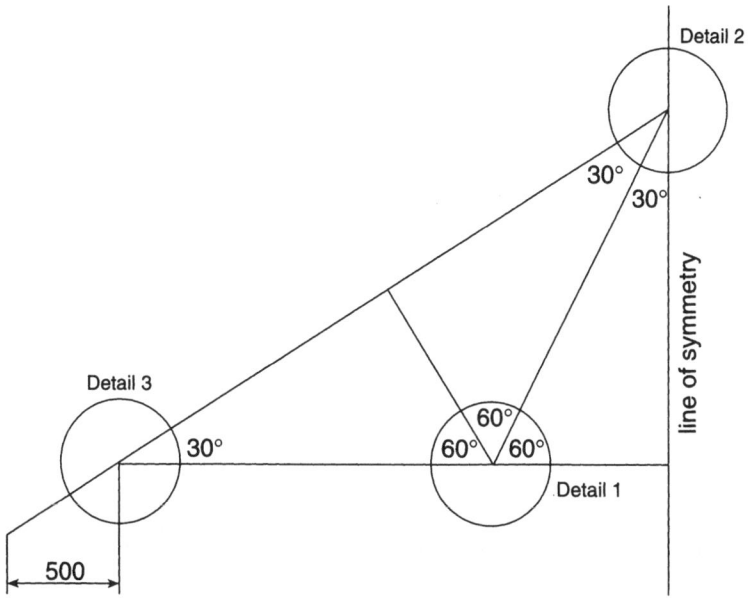

Fig 9.22 Diagram of steel roof truss

Activity 9.2

(Answers not included)

Refer to Fig 9.23 on page 174, and draw the typical bolted connection for the steel roof truss and H-section column to a scale of 1:5.

You can assume the following:
- Rafter: 80 mm × 80 mm × 10 mm angle (5 × 20 mm diameter bolts);
- Tie beam: 100 mm × 100 mm × 10 mm angle (3 × 24 mm diameter bolts); and
- Column: 254 mm × 254 mm × 107 kg.

H-section
1. Use 10 mm gusset and cap plate.
2. Use 2 × 80 mm × 80 mm × 10 mm shoe cleats, welded to the gusset plate and bolted to the cap plate by using 3 × 20 mm diameter bolts per cleat.
3. Use 6 mm fillet welds.
4. Fully dimension this detail drawing.
5. Note that the cap plate should be welded to the column.

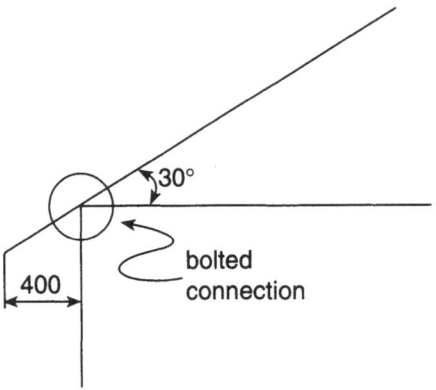

Fig 9.23 Typical bolted connection for steel roof truss

Activity 9.3

(Answers not included)

Fig 9.24 shows the layout of a steel roof truss that relies on equal angles. Draw the typical bolted connection for Detail 1 to a scale of 1:5. Show all necessary dimensions to enable trouble-free manufacture of this apex connection. Use a 10 mm thick gusset plate.

- Rafter: 100 mm × 100 mm × 12 mm (3 × 24 mm diameter bolts per member);
- King post: 100 mm × 100 mm × 12 mm (3 × 20 mm diameter bolts per member); and
- Diagonals: 80 mm × 80 mm × 10 mm (2 × 20 mm diameter bolts per member).

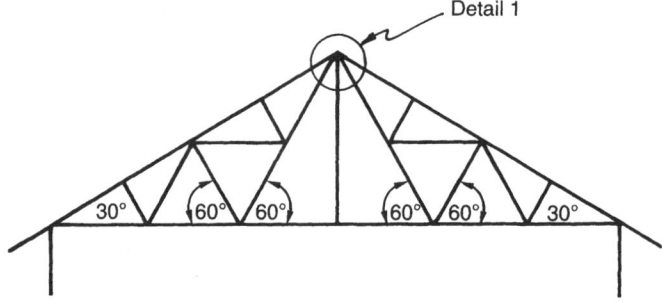

Fig 9.24 Steel roof truss consisting of equal angles

Activity 9.4

(Answers not included)

Fig 9.25 on page 176 shows an elevation of a portal frame structure for a factory. Take note of the following properties:

- There are seven portals at 7 m, centre to centre.
- The ridge and eaves haunches are cut from 203 mm × 133 mm × 30 kg/m parallel flange I-section, the end plates are 16 mm thick, and continuous 8 mm fillet welds are used.
- The connections are made using 8 × M20 grade 4.6 bolts through the end plate at each haunch.

1. Draw an elevation of one half of the portal frame, showing all dimensions and details. Use a scale of 1:20.
2. Draw the details of the haunch connections at B. Show an elevation and section to indicate the detail of the bolting to the column. Use a scale of 1:10.
3. Draw the details of the base plate-to-column connection.
   - Make use of 8 mm fillet welds.
   - The positions of the holes for the four M25 grade 4.6 holding down bolts.
   - The edge distance for the bolts is 60 mm.
   - Use a suitable scale.
4. Draw the reinforcing details of a footing. The reinforcing is: Y16 — 100, shape 34, both ways, alternate bars reversed. Use a suitable scale.
5. Using a suitable scale, draw a fully dimensioned and annotated grid layout for the columns and footings of the above factory.

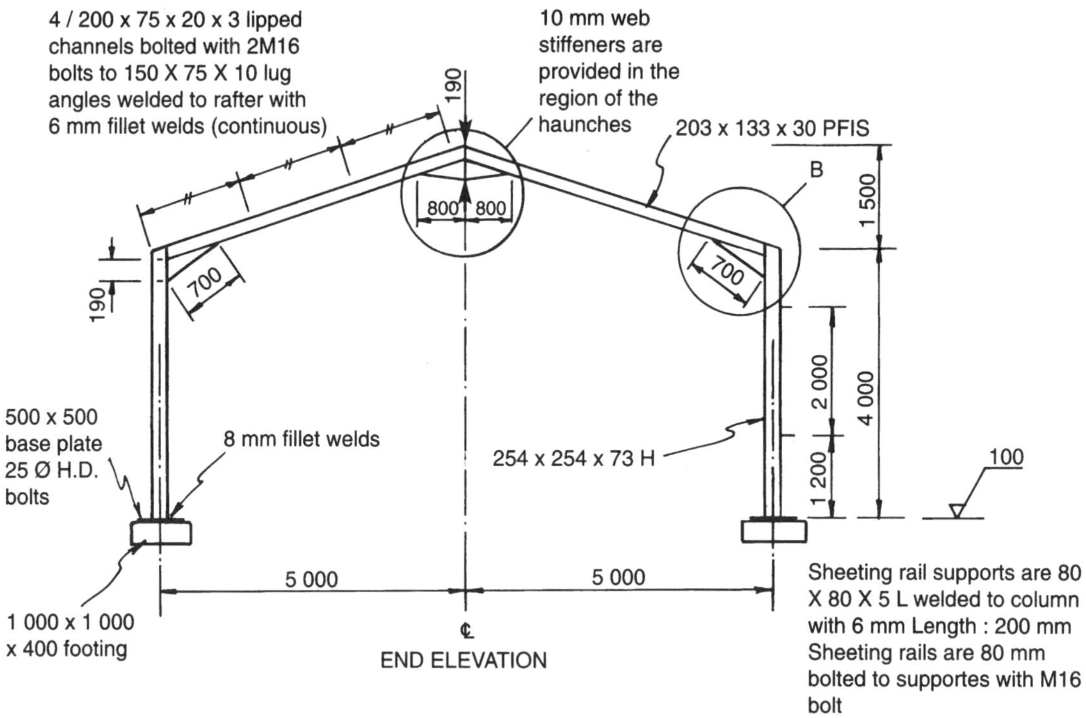

Drawing for Civil Engineering

Fig 9.25 Portal frame structure

Activity 9.5

(Answers not included)

Before you attempt this activity, you need to do some research on bracing by visiting the library and/or some engineering companies. Fig 9.26 on page 177 is a side elevation of a factory showing the cross bracing between two portal frames in Activity 9.4. Each cross brace consists of an 80 mm × 80 mm × 8 mm angle. These cross braces are connected together at their crossing point by means of an M20 bolt and bolted to gussets at each end using 2 × M16 bolts.

At the base end of the cross brace the gusset is welded to the centre of the column web and to the base plate, using continuous 6 mm fillet welds.

1. Draw an elevation of the cross bracing and columns shown, giving the details of the angles and dimensions. Use a suitable scale.
   - End distance = 30 mm;
   - Pitch = 50 mm; and
   - The bolting is done on site.

Module 2: Unit 9

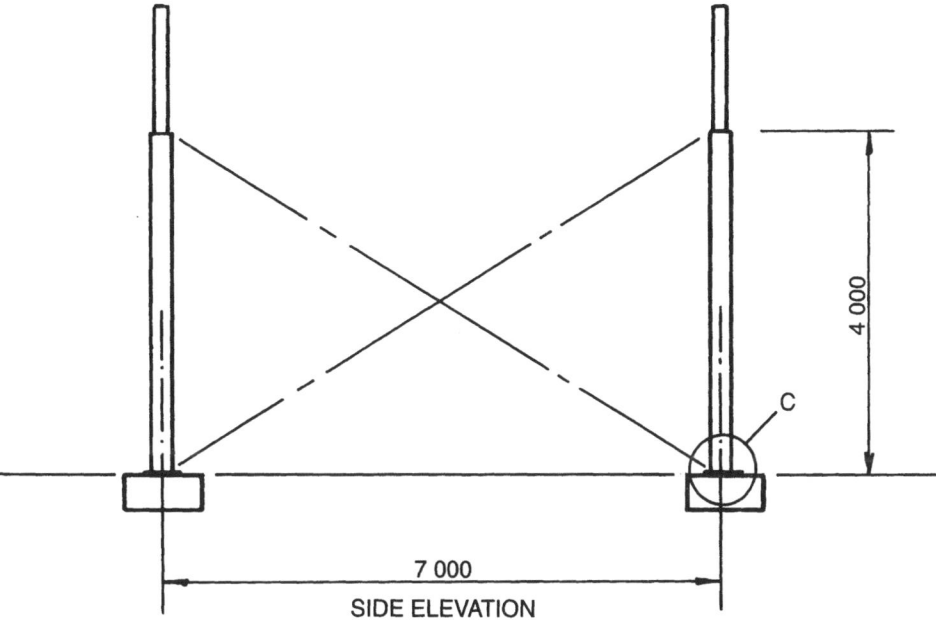

Fig 9.26 Cross bracing between two portal frames

2. Draw the details of the gusseted joint at C, showing a portion of the angle, the bolts, welds and the centroidal line. Use a suitable scale.
   - End distance = 30 mm;
   - Pitch = 50 mm; and
   - The bolting is done on site.

Self-evaluation

1. What is a roof construction and what is its main function?
2. Briefly describe the different types of roof structures.

SELF-EVALUATION ANSWERS
1. A roof construction consists of members constructed in such a way that they will be able to carry the roof covering material.
2. a. Roof truss: A plane frame consisting of sloping rafters that meet at the ridge, a main tie that connects the feet of the rafters and the internal bracing members.
   b. Lattice girder: A plane frame of open web construction, having parallel chords and internal web bracing members.
   c. Portal frame: A type of arch construction in which the roof member is joined rigidly at the eaves to the stanchion.

module **3**

Surveying

The photograph shows surveying in action.

Photo credit: Cascade Surveying and Engineering, Inc.

Introduction to surveying drawing for civil engineering

Module outcomes

After studying this module, you should be able to:
- draw a grid system to scale, enter the co-ordinate values correctly on to the grid and indicate true north;
- plot co-ordinated points on the grid system;
- plot spot-heights;
- interpolate and draw contours;
- plot relevant detail using the correct symbols;
- correctly position a structure on a contour plan;
- draw a longitudinal section of a road, pipeline and a dam wall;
- indicate on the longitudinal section the chainages, ground levels, formation levels and slopes of the structure;
- determine the cross-sectional areas by Trapezoidal rule;
- calculate volumes using the end area rule and plot mass haul diagrams;
- determine the freehaul volume and freehaul given the freehaul distance;

Drawing for Civil Engineering

- determine the overhaul volume and overhaul; and
- determine the volumes of borrow or spoil.

Module 3 consists of Units 10, 11 and 12.

 The words below are printed in bold the first time they are used in the text.

Average haul distance: The distance from the centre of gravity of the cut to the centre of gravity of the fill.
Cadastral data: Data that include boundary lines of properties and other relevant information.
Contours: Lines of equal elevation.
Cross-section: A section that runs perpendicular to the longitudinal section.
End area method: Volume $(m^3) = \dfrac{(A_1 + A_2)}{2} \times$ perpendicular distance between areas A_1 and A_2
Equator: An imaginary circle whose plane is perpendicular to the earth's axis, at equal distance from the North and South Poles.
Freehaul: A specified distance used in mass haul diagrams.
Grid lines: Lines parallel to the equator (X grid lines) and the central meridian (Y grid lines).
Haul: The volume moved over the specified distance.
Haul distance: The distance from the working face of an excavation to the tipping end of the embankment.
Interpolate: To find, mathematically, an accurate value between two known values.
Limit of economical haul: The maximum distance determined by the contractor. (The maximum overhaul distance plus the freehaul distance.)
Longitudinal section: A section that runs horizontally along the centre line of a proposed engineering project.
Meridians: Circles of constant longitude, passing through the poles.
Origin: A fixed imaginary point from which co-ordinates are measured.
Overhaul: A volume moved beyond the freehaul distance.
Spot-height: The reduced level of each reading taken by means of a theodolite.
Surveying: The art of taking measurements on the surface of the earth either in the horizontal or vertical plane.
Topographical plan: A map that represents the natural and artificial features of the earth's surface, for example hills and rivers, roads and houses and so on.
Trapezoidal rule: $\text{Area}[m^2] = \dfrac{(\text{Sum of parallel sides})}{2} \times$ perpendicular height

Unit 10 Introduction to surveying (refer to S1 notes)

10.1 Introduction

Remember that we covered this section (including the practical aspects) in the *S1 Survey and Drawing* course. Therefore, most of this section should be revision for you.

In **surveying**, we take measurements on the surface of the earth either in the horizontal plane or in the vertical plane, by using instruments such as a level and a theodolite. The practical aspects of surveying were covered in *S1*. In civil engineering and in surveying it is important to produce accurate plans of sites on which structures are to be erected. On the one hand, as a *survey technician*, you would be involved in producing topographical plans of the site, and in providing co-ordinated points for the construction of the structure involved. On the other hand, as an *engineering technician*, you would be involved in producing smaller **topographical plans**, and in setting out and monitoring the construction of a project. This unit deals with how to plot **grid lines**, how to survey stations and **spot-heights** and how to **interpolate** and draw **contours**.

In planning line structures such as roads, dam walls and pipelines, it is essential to know the position of the structure relative to the ground level. This will help you to calculate the volumes of cut and fill material between any two points on the line structure, correctly position the structure to attain the required slopes, and so on.

We use contour plans to draw the **longitudinal section**. Once this is done, we calculate the volumes of earth to be moved and we draw a mass haul diagram. We use the mass haul diagram to select the earth moving equipment and to do the necessary cost calculations. This will be dealt with in the course called *Documentation*.

10.2 Grid lines (revision)

To find the position of any point in plan on the surface of the earth, it is necessary to know the co-ordinates for that point you are trying to position. (Make sure that you know what co-ordinates are.) We calculate a point by measuring the perpendicular distance from each of the two co-ordinate axes, the intersection of which is called the **origin**. In a plan the two co-ordinates of any point are usually referenced by Y and X values.

Remember that the Y co-ordinate is written first, followed by the X co-ordinate, for example:
Y − 19 426,13 X + 314 006,07

In the South African co-ordinate system, we measure the Y co-ordinate positive (+) to the west of any origin, and negative (−) to the east of that origin. We measure the X co-ordinate positive (+) to the south of the **equator**, and negative (–) to the north of the equator.

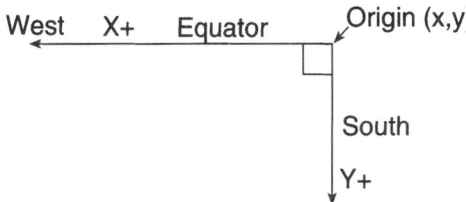

Fig 10.1 Grid line convention

Grid lines are imaginary lines drawn parallel to the Y and X axes, and they are always drawn as straight lines perpendicular to one another. The grids are usually drawn with north pointing to the top of the page. Remember that in the polar co-ordinate system, 0° is south, increasing in the clockwise direction. See Fig 10.2 to refresh your memory. The grid (or rectangular co-ordinate system) serves the following purposes:
- It serves as the survey co-ordinate system.
- It enables co-ordinated points to be plotted accurately.
- Rough co-ordinates of points fixed by other means may be scaled off.
- It allows distances and directions to be plotted or scaled easily.

Let us fix the position of point P_1 for the known angle θ, and distance d from point c.

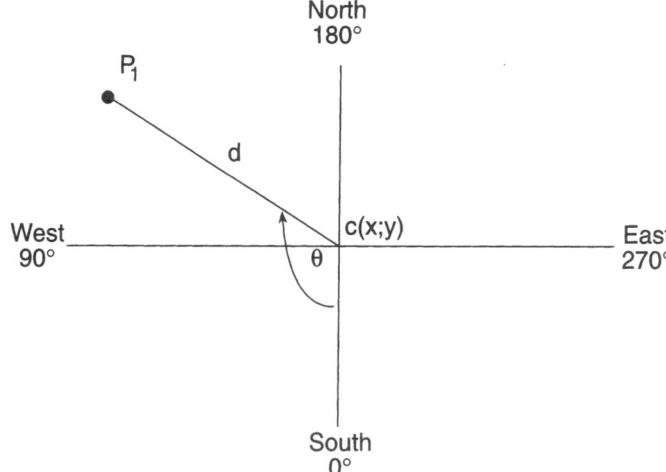

Fig 10.2 Clockwise direction from south

Module 3: Unit 10

10.3 Plotting spot-heights (revision)

In practice we use a tacheometer to level traverse pegs. We measure the distances, the vertical angles and the horizontal angles. Thereafter, as you were taught in *S1*, reduction of the heights is carried out. In the drawing room we start plotting spot-heights. We use a protractor for angle θ and a scale for distance *d* (see Fig 10.2 on page 182). A dot (•) is used to represent the position, the decimal comma and the label of the spot-height. In Fig 10.2, the spot-height is represented by P_1.

10.4 Contouring (revision)

Do you remember what a contour symbolises? To remind you: contours are lines of equal elevation. A general rule is to presume that the ground between spot-shots slopes evenly, so that the position of the contours is determined by interpolation.

Accurate interpolation can only be done if the direct distance between the spot-heights is not very big.

10.4.1 Draw in contours

How do you interpolate contours? The most accurate method is as follows:
1. Scale the distance between the spot-shots that are going to be used to interpolate the contour.
2. Calculate the difference in height between the spot-shots.
3. Calculate the positions of the contours between the spot-heights.

Remember the mathematical method that we use for interpolation (proportionality). Fig 10.3 helps to clarify this concept.

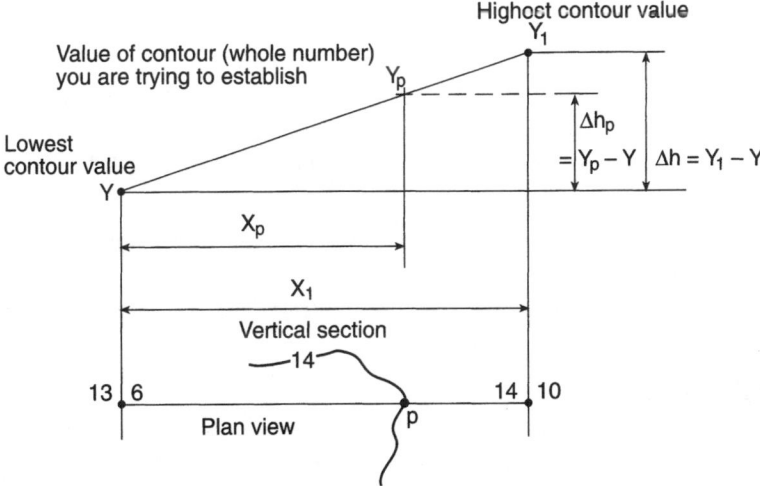

Fig 10.3 Method of interpolation

183

Drawing for Civil Engineering

$$\frac{X_p}{\Delta h_p} = \frac{X_1}{\Delta h} \Rightarrow X_p = \frac{X_1 \Delta h_p}{\Delta h}$$

This corresponds to the formula $X_p = \dfrac{X_1(Y_p - Y)}{\Delta h} = \dfrac{X_1(Y_p - Y)}{Y_1 - Y}$

To clarify this, see Fig 10.4. In the plan view, let us consider $Y = 13{,}60$, $Y_1 = 14{,}10$, $Y_p = 14{,}00$ and the scaled distance $X_1 = 35\,\text{mm}$, where Y values are vertical distances, and X values are horizontal distances.

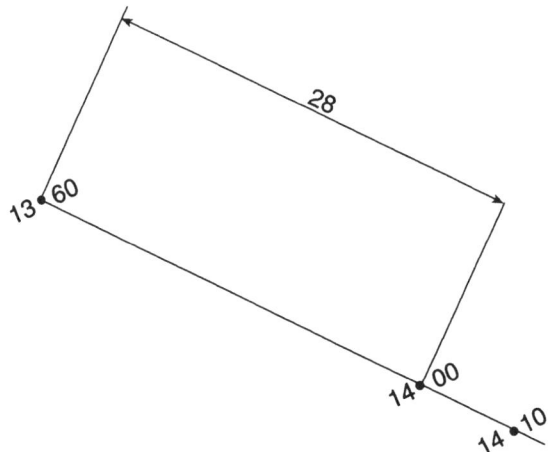

Fig 10.4 Plotting of spot-heights

In our example, this means that if we measure 28 mm on the straight line between 13,60 and 14,10 and 13,60 m from spot-height we can find the 14 m contour height.

$$X_{14} = \frac{35(14{,}0 - 13{,}6)}{14{,}1 - 13{,}6} = \frac{35(0{,}4)}{0{,}5} = 28\,\text{mm}$$

This process is continued until all the values for the 14 m contours are found. Then we combine these spots by means of a smooth, curved line to represent the contour of 14 m. Remember that practice makes perfect, and contouring can be fascinating. *Now calculate the position of the 14 m contour between spot-heights 13,65 and 14,30.*

Contours are always indicated as whole numbers, for example 50 m, 65 m and so on.

Can you see why we do not interpolate between the spot-heights 14,10 and 14,30?

Module 3: Unit 10

10.4.2 Contour values

Contour values appear on either side of the contour line as shown in Fig 10.5 or Fig 10.6.

Fig 10.5 Open contours

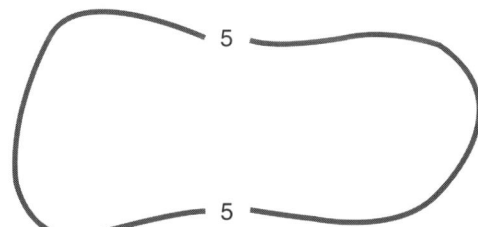

Fig 10.6 Closed contours

When drawing a profile (section), contour values will be upright.

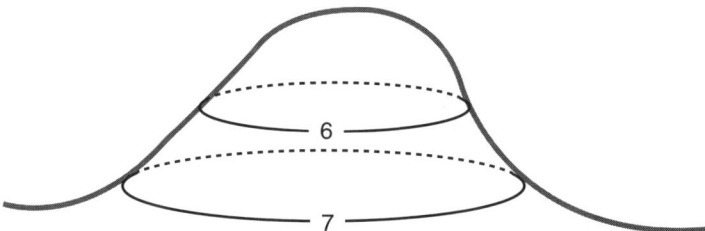

Fig 10.7 Contours in section

When drawing a plan, you can use one of the two methods shown in Fig 10.5 and Fig 10.6, as shown in Fig 10.8 on page 186.

185

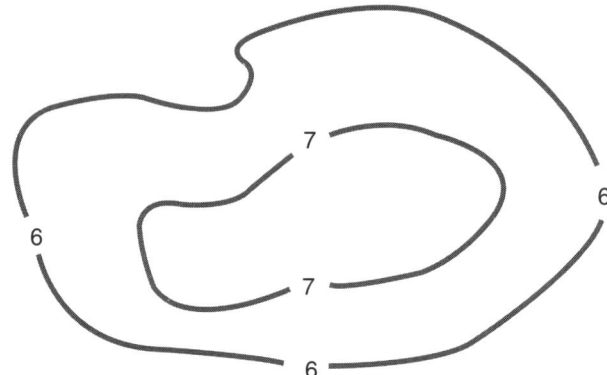

Fig 10.8 Plan view of closed contours

10.4.3 Contour characteristics

- Closely spaced contours indicate steep slopes.
- Widely spaced contours indicate moderate slopes.
- Contours must be labelled to give the elevation value.
- Contours are not shown passing through buildings.
- Contours crossing a man-made surface (roads, railroads) will be straight parallel lines as they cross the facility.
- Since contours join points of equal elevation, they cannot cross each other.
- Contours deflect uphill at valley lines and downhill at ridge lines; line crossings are perpendicular.
- Contour lines tend to be parallel to each other on uniform slopes.

10.4.4 Physical features

As soon as spot-heights, which were taken onto natural or artificial features, are plotted, they must be joined with the other spot-heights for the same features. Written spot-heights will only clutter up the feature. Making sketches and notes for every spot-height makes this task easy.

10.4.5 Conventional symbols

A thorough knowledge of the symbols used on Cadastral plans is necessary. Study Table 10.1 on pages 187 to 188 to see the standard symbols that are currently in use in survey maps.

Module 3: Unit 10

Table 10.1 Standard symbols used in survey maps

ROADS, ETC.	
Tracks & Paths (including concrete pavement paths, etc.)	‑ ‑ ‑ ‑ ‑ ‑ ‑ ‑ ‑ ‑ ‑ ‑ ‑ ‑ ‑ ‑ State surface:
Footpaths	= = = = = Freehand
Hardened roads	— — — — — Tar — —
Bridges & Culverts	⌐_/ ⌐‾\ State material of Bridge or Culvert. Bridge to be drawn as sketch or in detail.
Kerb & Channel	Channel / Kerb
Gutter Bridge	‑ ‑ ‑ ∠ ‑ ‑ ‑
Kerb & Gutter Dished	= ⌐ = =
Railways	▬▬▭▬▬▭▬▬
Cane Tracks or Narrow Gauge	+++++++++++++
Letter Boxes, Bollards, Fire Alarms, Petrol Pumps, Robots, etc.	Fire Alarm ○ — Description in full. Actual shapes may be shown.
All Street Signs or Notices	✕

PLAYGROUNDS

Playgrounds Swimming pools, etc.	┆ Tennis ┆

UNDERGROUND AND OVERHEAD SERVICES	
Water Mains	100 mm Water Main ——W—— If exposed, write 'Exposed' next to main at the entrance and exit.
Water Manhole	☐ V
Water Valve	○ M
Water Meter	○ H
Fire Hydrant	FP ○
Fire Plug	○
Reservoirs Filtration Beds, etc.	Reservoirs (covered) — State 'Open' or 'covered'
Water Towers	T — To scale of base
Sewer Main	150 mm Sewer Main — · — · — · — If necessary write 'Exposed'.
Sewer Manhole	☐ S
Stormwater Drain	300 mm Sewer Main —SW— · · · —
Stormwater Manhole	☐
Stormwater Inlet	= =
Stormwater Catch Pit	CP ☐
Stormwater Culvert over 1 m	Culvert 2,5 m wide
Telephone Wires or Cables	— · — · — · — State if U/G. Tel
Telephone Manhole	☐ J
Telephone Joint Box	○
Telephone Pole	◐
Electric Cables or Wires	— — — State if U/G.
Electric Manhole	☐
Electric Pole	●
Electric Switch Pillar	⊠
Stay Pole	○
Railway Electrification Masts	☐
Pylons	— — ⊠ — — To scale
Substations	SS — To scale
Oil Pipe Line	100 mm Oil Pipe Line — · · — · · —
Oil storage tanks	○ To scale
Invert levels	(27-73)

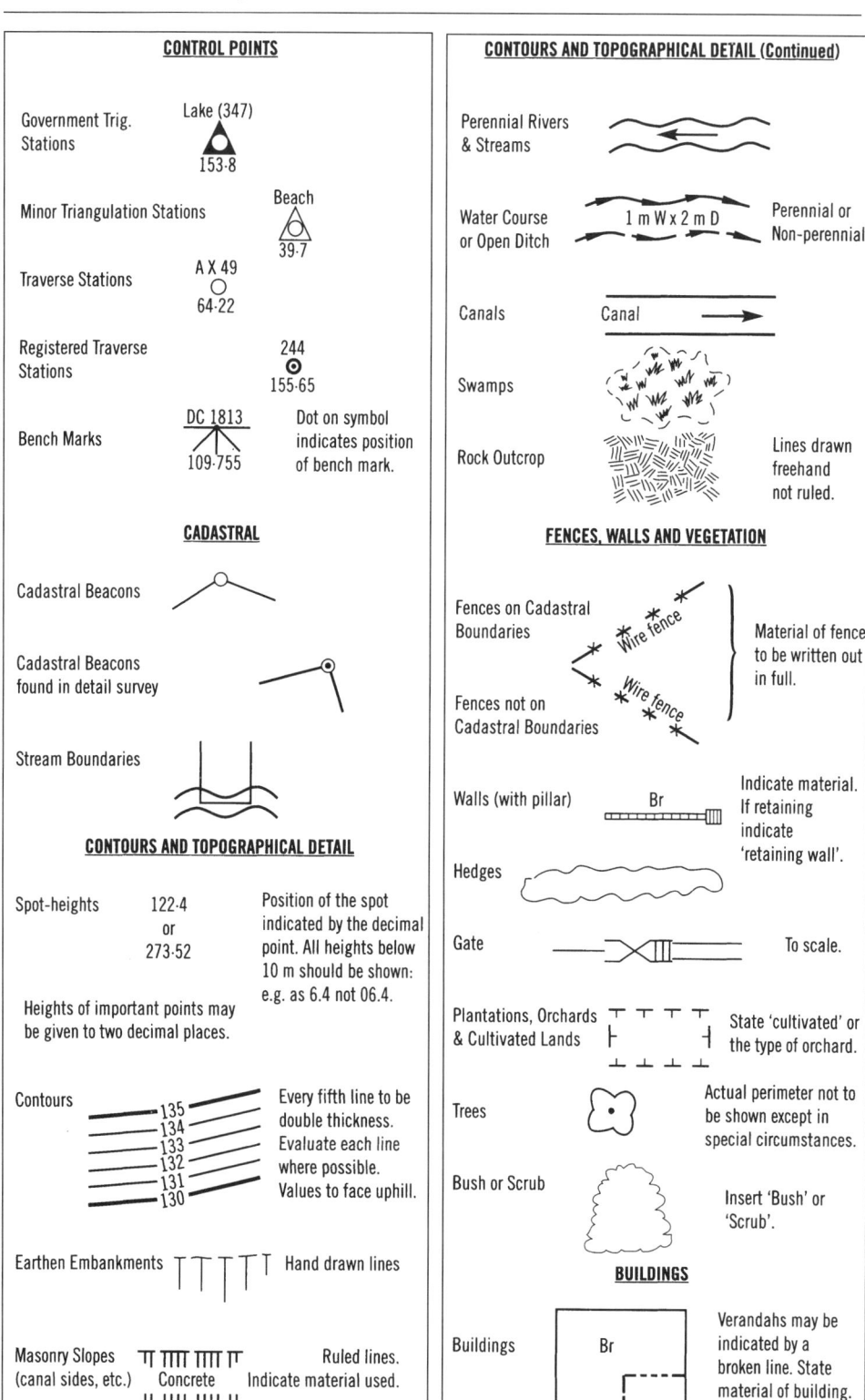

Module 3: Unit 10

10.5 Step-by-step method of drawing grids and contours

10.5.1 Necessities on any plan

Use the following points as a checklist when drawing or looking at any plan.

Grid lines

Grid lines are thin, dark lines, parallel to the equator (X grid lines) and perpendicular to the central **meridian** (Y grid lines).

Grid values

See 10.5.2 Method 2, point 7, for clarity.
- X grid values decrease towards the equator.
- Y grid values decrease towards the central meridian.
- Show full grid values (with positive or negative signs) on both ends of the X and Y grid lines.
- Print grid values parallel to each grid line.

North point
- Draw parallel to the Y grid lines.
- The north point points in the direction in which the X grid values decrease (that is, towards the equator for the southern hemisphere).
- Should be visible and neat.

Scale of plan
- Print in the title block or below the drawing.
- Natural scale is written as a ratio, for example scale 1:1 000.

Once everything on the checklist appears on the plan, you are ready to continue with the next stages.

10.5.2 Orientation of the plot and plotting grid lines

How will you ensure that the entire survey fits centrally onto the drawing? There are essentially three methods of doing this.

Method 1: Co-ordinates centred, grid lines parallel to the edges of the drawing sheet

Given co-ordinates:

Station	Y co-ordinate	X co-ordinate
A	+5 763,21	+6 777,86
B	+2 314,68	+3 576,84
C	+3 642,28	+9 876,57

189

1. Find the maximum and minimum Y and X:
   [Y 5 763,21 and 2 314,68; X 9 876,57 and 3 576,84]
2. Round the maximum values up to the next 100: [Y 5 800; X 9 900]
3. Round the minimum values down to the next 100:
   [Y 2 300; X 3 500]
4. What is the available space between the maximum and minimum of rounded Y and X values?

   Largest   Y = +5 800   Largest   X = + 9 900
   Smallest   Y = +2 300   Smallest   X = + 3 500
              3 500 units   6 400 units

The highest (largest) values and the smallest values are within the limits of the given co-ordinates.

5. We plot the bigger difference (6 400) on the longer side of the drawing sheet.

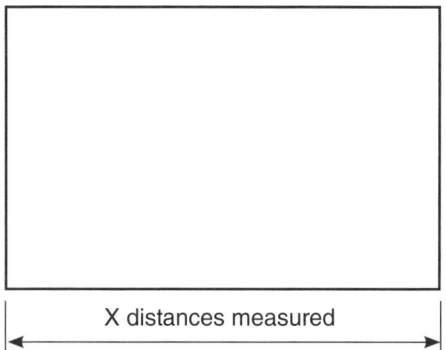

Fig 10.9 X distance

6. Use a scale ruler and choose a scale, so that 6 400 units fit across the sheet. Test that using the same scale, the 3 500 units will fit over the width of the sheet. If they do not, choose a smaller scale.
7. Draw light construction lines which define the centre of the paper. (Drawing diagonals is the easiest method of achieving this.)

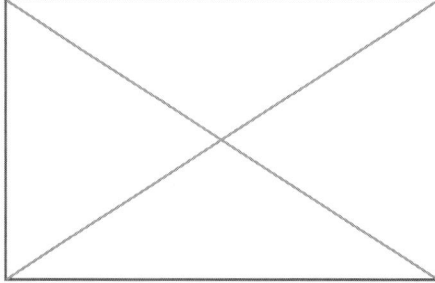

Fig 10.10 Finding centre

8. From the centre point, measure half the number of units to either side, at the chosen scale.

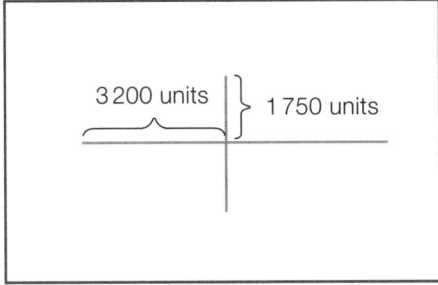

Fig 10.11 Measuring units in both directions

$$\frac{6400}{2} = 3200 \text{ units}$$

$$\frac{3500}{2} = 1750 \text{ units}$$

9. Insert the maximum and minimum grid values in pencil.
10. Determine the north direction and reposition your drawing sheet so that N points up.

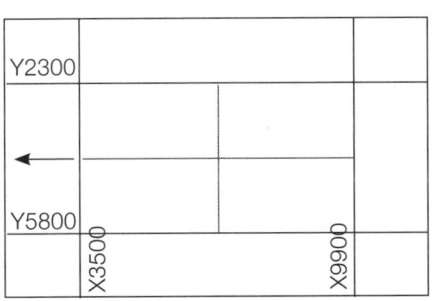

Fig 10.12 North direction

11. Now recheck your Y and X grid values.

Method 2: Co-ordinates centred, grid lines determined by direction and distance

Given co-ordinates: any two co-ordinates are used; the others are used as a check).

Station	Y co-ordinate	X co-ordinate
A	+5 763,21	+6 777,86
B	+2 314,68	+3 576,84
C	+3 642,28	+9 876,57

1. Here we choose co-ordinates A and B. Do a join, that is:
   Direction A to B = 227° 07' 54"
   Distance = 4 705,198 m

 Make sure that you still remember how to calculate joins!

2. Determine the centre of the sheet using light construction lines.

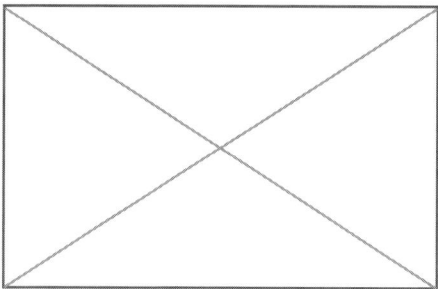

Fig 10.13 Finding centre

3. From the centre point, measure half the join distance along the longest side using a scale which is either given or chosen to fit:

$$\frac{4\,705,198}{2} = 2\,352,599$$

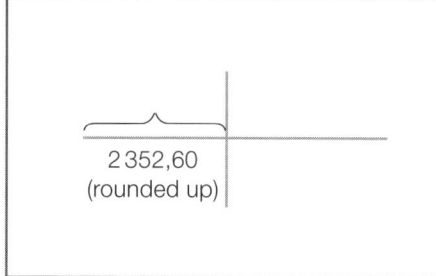

Fig 10.14 Measuring units

Module 3: Unit 10

4. From this mark, measure the full distance along the construction line. These two marks represent stations A and B. Some use a circle with a cross through it or a circle with a dot inside to indicate a survey station.

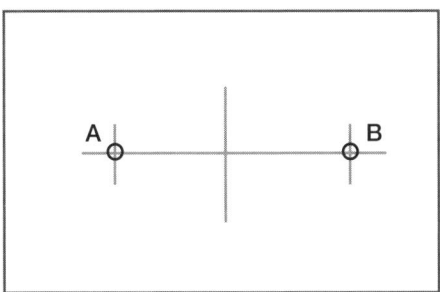

Fig 10.15 Plotting stations A and B

5. Place your protractor over A, orientate it so that the join direction (227° 08') is in line with the construction line, and carefully mark the 0°, 90°, 180° and 270° points on the sheet.

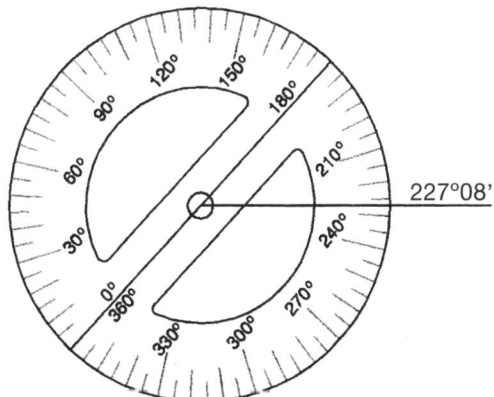

Fig 10.16 Orientation

6. Use construction lines to join the marks (corresponding to the 0°, 90°, 180° and 270° points) and extend them to the edges of the paper.

- These two lines are parallel to the grid lines.
- The 0° + 180° line indicates the direction of the north arrow (decreasing X value).

7. You must establish in your mind where the equator is and where the central meridian is. This tells you in which direction the grid values increase (see Fig. 10.17 on page 194).

193

Drawing for Civil Engineering

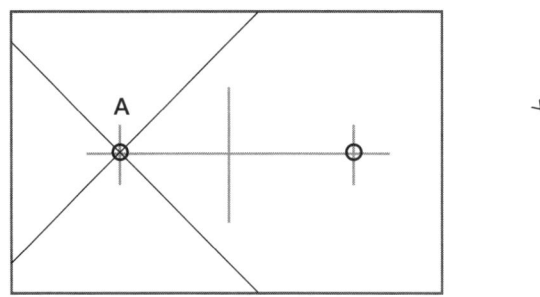
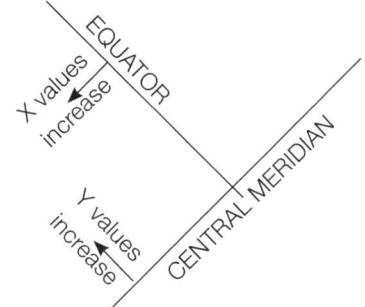

Fig 10.17 Direction to which grid values increse

8. Determine the even, whole grid values smaller than the co-ordinates of A (5 600 Y; 6 600 X). Subtract these from the co-ordinate of A, that is:

   +5 763,21   + 6 777,86
   − 5 600,00 Y and   − 6 600,00 X
   = 163,21   = 177,86

9. Measure 163,21 units from the construction line passing through A, considering the sign of Y. Repeat for X.

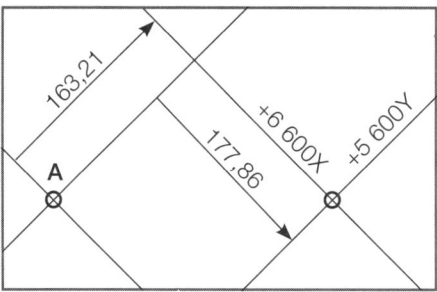

Fig 10.18 Drawing grids X and Y

10. Continue measuring away from these grid lines for the other grid lines.
11. Check that the co-ordinates of B do indeed plot in the correct position.
12. As soon as all the grid lines and values are on the plan, plot station C relative to your grid lines.

Method 3: Accurate grid lines onto a sheet

The first stage in plotting co-ordinates is to establish a co-ordinate grid. We draw grid lines at specific intervals, for example 100 m, 200 m or 300 m, parallel to the equator (for X grid lines) and parallel to the central meridian (for Y grid lines) to form a pattern of squares. We then plot the stations in relation to the grid. In this method T-squares and set-squares should not be used.

1. From each corner of the drawing sheet, draw two diagonals.

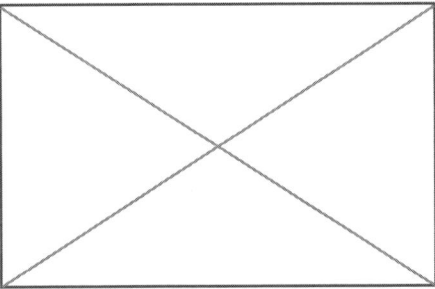

Fig 10.19

2. Use an extension arm on your compass or a beam compass to mark equal distances on each diagonal, away from the central point. (Set your compass at any distance.)

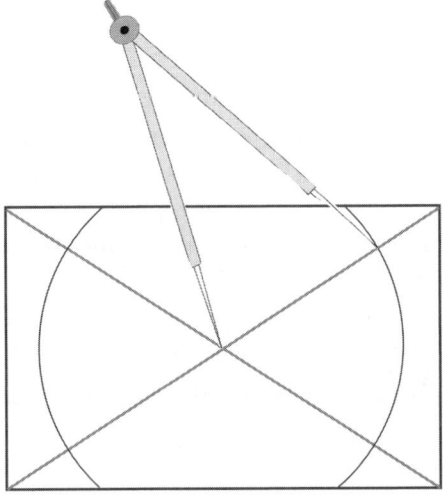

Fig 10.20

3. Join the marks you have made and form a rectangle, still using faint lines.

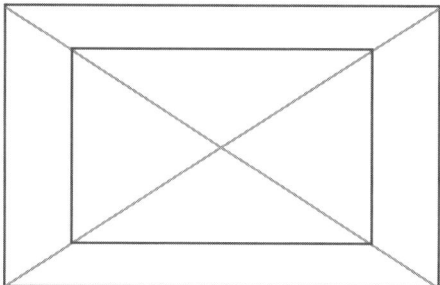

Fig 10.21

4. Measure distances K (= grid interval) along lines AD and BC at your chosen scale, and join lines AD and BC to form the Y grid line. Similarly, measure distance L (= 2 × grid interval) along AD and BC. Measure distance from line AB, as reference line, to eliminate accumulating errors.

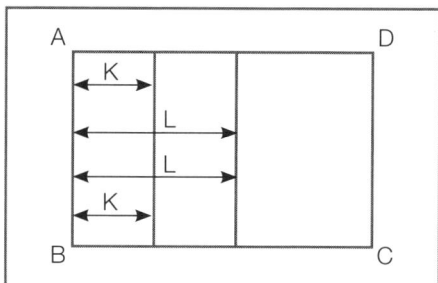

Fig 10.22

5. Establish the horizontal grid lines in a similar manner.
6. We now enter grid values on both ends of the lines (left and right of the drawing sheet for X values, and bottom and top for Y values). These must be full values, multiple(s) of 50, which may include the constants. Remember that these values should be inserted parallel to the lines.
7. Plot co-ordinates relative to these grid lines. After plotting all the co-ordinates, join distances between stations are used to check the plotting of the stations.

Note that you cannot continue with the plotting of the **Cadastral data** or spot-heights unless the survey stations have been checked.

10.5.3 Plotting sequence

1. Plot co-ordinates of survey stations

Station	Y co-ordinate	X co-ordinate
A	+20 972,00	+2 184 485,00
Elevation 1 267,432		
B	+20 812,96	+2 184 546,87
Elevation 1 264,326		

Measure 172 m from the +20 800 Y grid line, as shown in Fig 10.23. Line up your scale ruler on these two marks. Now measure 85 m from the +2 184 400 X grid line. The position of survey station A is at the intersection of these two measured distances. The position of survey station B can be found in a similar way. See if you can position survey station B on your own.

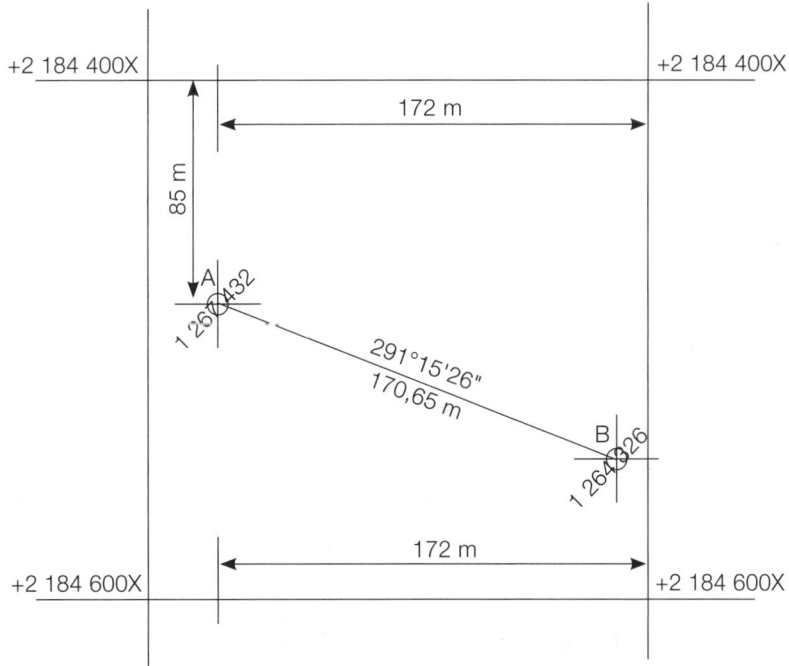

Fig 10.23

2. Plot spot-heights

You only need four pieces of information from the Tacheometry field book:
1. horizontal distance;
2. direction;
3. elevation (height at that spot); and
4. the number or name of the spot-height in the remarks column.

Spot-heights can be plotted using the following two methods:
a. Protractor or a scale ruler.
   i. Orientate the protractor at the station where the spot-heights were taken.
   ii. Draw faint lines through point A, parallel to the grid lines.
   iii. With 180° pointing north, mark the directions of the spot-shots on the edge of the protractor.
   iv. From the point A, measure the horizontal distance to the spot-height in question, and mark it with a dot. Print the level of the spot-height.

The dot represents the position as well as the decimal point of the spot-height, for example, 15•64.

b. Plotting protractor with attached scale ruler. This is a special case, which uses a special device.
   i. Join AB to obtain direction AB.
   ii. Join station A and B with a faint line.
   iii. Position centre plotting protractor on A and line up the attached scale ruler on the faint line between A and B.
   iv. Mark exactly on the drawing sheet where 0° is on the plotting protractor. Check the join distance with the distance on the scale ruler — this checks that your previous plotting was correct, and that you are using the correct scale.
   v. Draw a line through point A and the mark you have made corresponding with 0°.
   vi. This line represents the reference line from which the direction to station B is measured.
   vii. Use this line as a reference line to measure all the other directions and distances to each spot-shot. Mark the point with a dot.

3. Interpolate and draw the contours

Refer to section 10.4 of this unit, which discusses contours in some detail.

4. Plot Cadastral data

Cadastral data is data that indicates among other things, property boundaries, servitudes and the like. Table 10.1 on pages 187 to 188 shows the standard symbols that are presently in use.

10.6 Summary

This section was based on work that was covered in *S1 Drawing*. It mainly involved conveying the results of the practical work that was covered in *S1* on to a plan. You should acquaint yourself with the procedure of drawing grid lines and contour plans.

SOLVED EXAMPLE

EXAMPLE 10.1

Let us go through a typical example to demonstrate the whole plotting sequence. Remember that we are not going through the practical procedure of surveying the site. We start our example after the site has been surveyed and the levels have been reduced.

You are requested to design a sewer pipeline. There is no contour plan of the site available and therefore you have to draw a contour plan of the site.

Part A: The contour plan
The survey results are as follows:

Station	Spot-height	Co-ordinates Y	System Lo 27 X
Constants		−80 000,00	+ 200 000,00
A	9,35	−1 266,00	+ 4 641,50
B	6,60	−1 182,50	+ 4 487,80
C	8,58	−1 421,20	+ 4 452,50

Plot these survey stations to a scale of 1:1000 m. Draw your grid lines at 50 m intervals. The following spot-heights, distances and directions are given from these survey stations:

From survey station A:

Spot-height	Horizontal distance in metres	Direction in degrees
8,42	45,80	28,20
7,42	110,60	72,60
5,56	96,50	301,80
5,82	112,70	257,00
6,11	94,90	275,00
4,62	171,00	273,50
8,24	51,40	206,50

From survey station B:

Spot-height	Horizontal distance in metres	Direction in degrees
7,85	129,20	340,00
5,75	67,50	316,00
6,45	74,00	359,20
4,25	66,40	72,40
6,20	126,50	31,50
7,92	57,00	267,70
7,68	56,00	188,20

From survey station C:

Spot-height	Horizontal distance in metres	Direction in degrees
7,50	140,00	52,40
8,10	72,00	51,50
6,54	119,50	16,80
6,05	123,50	348,00
10,10	72,20	131,00
9,85	122,80	103,60

Plot the spot-heights according to the above distances and directions and interpolate the contours at 1 m intervals. Draw the contours neatly, and indicate their height on your plan. Neatly print the scale below the drawing.

Solution

See Fig 10.24 in the pocket at the back of the book.

Module 3: Unit 10

Activity 10.1

(Answers not included)

The following co-ordinates refer to the survey stations A, B and C:

Station	Spot-height	Co-ordinates Y	Co-ordinates X
Constants		− 75 000,00	+ 220 000,00
A	34,71	− 2 029,00	+ 2 197,00
B	32,15	− 2 121,20	+ 2 062,50
C	35,10	− 2 338,00	+ 2 009,20

Plot these survey stations to a scale of 1:1000. Draw grid lines at intervals of 50 m. The following spot-heights, distances and directions are given from these survey stations:

From survey station A:

Spot-height	Horizontal distance in metres	Direction in degrees
34,05	68,80	156,60
35,08	34,00	157,80
32,61	140,50	266,40
35,22	26,00	281,00
33,10	91,80	192,60
35,31	30,70	220,80

From survey station B:

Spot-height	Horizontal distance in metres	Direction in degrees
31,20	52,00	235,70
30,58	111,00	291,00
31,72	61,00	294,50
33,45	74,20	358,50
31,50	122,70	317,30
32,18	88,00	322,00
33,32	52,30	13,50

From survey station C:

Spot-height	Horizontal distance in metres	Distance in degrees
29,92	116,00	78,00
30,00	91,00	71,50
29,41	111,00	64,20
35,95	112,50	23,50
35,12	63,70	34,00
32,30	124,80	35,30
32,08	87,20	47,80

Plot the spot-heights according to the above distances and interpolate the contours at 1 m intervals. Draw the contours neatly, and indicate their heights on your plan. Neatly print the title and scale below the drawing, and show a north point.

Self-evaluation

1. What do you understand by surveying?
2. Why is the plotting of surveying data so important for the:
   a. civil engineering technician?
   b. survey technician?
3. Explain the use of co-ordinates.
4. Define contours.
5. What is the meaning of interpolation?

SELF-EVALUATION ANSWERS

1. Surveying is the art of taking measurements in the horizontal and vertical plane in order to calculate directions and distances.
2. a. Plotting surveying data enables the civil engineering technician to produce topographical plans for the purpose of setting out and monitoring the construction of structures.
   b. Plotting surveying data enables the survey technician to produce topographical plans of sites and to provide co-ordinated points for the construction of structures.
3. A co-ordinate represents the Y and X values to plot any point in a two-dimensional way.
4. Contours are lines that indicate equal elevation of the natural topography.
5. Interpolation is the method of finding a certain contour value between two different heights.

Unit 11 Positioning of structures on a site plan

11.1 Introduction

Until now, we have dealt only with the representation of natural features on a plan. Now let us look at transferring information from a plan to the ground. The production of an accurate plan is of prime importance to the success of any construction project. This unit also deals with the positioning of a road, pipeline or dam wall on a contour plan, where it is necessary to achieve the correct gradients and alignment. You must familiarise yourself with the relevant specifications laid down by the Department of Transport and the Provincial Administration Standards, because those specifications are not discussed in this course.

11.2 Cut and fill

You should know that the natural ground surface is seldom suitable for engineering projects. The surface must be appropriately levelled for playing fields, parking grounds or industrial sites. Roads and railways must be properly graded, so that they are not too steep and do not consist of a series of humps and hollows. Trenches for drains must be dug to correct slopes so that the water will flow in the right direction, and so on. In all these cases, it is likely that we would have to remove ground from some places and add more ground in other places to bring the new ground surface to an exact, predetermined level. The removal of ground is known as 'cut', and the addition of ground is known as 'fill'. The level of the completed earthworks, before the addition of any surfacing material or special foundations, is known as the 'formation level' (final level).

11.3 The mass haul diagram (MHD)

11.3.1 General

Earthworks involves three operations, namely excavation, haulage and the formation of embankments (fill). Remember that the costs of excavation and embankment depend on the properties of the soil. The amount and cost of haulage depends on three things, namely the profile of the new road or structure in relation to the old ground line, the economic positioning of borrow pits and spoil heaps, and good judgement as to the direction of haul. We draw the mass haul diagram (very often directly below the longitudinal section of the surveyed centre line) after calculating the earthwork quantities.

So, what is the function of a mass haul diagram? It shows the accumulated volume at any point, and from it we can determine the following:

- distances over which cut and fill will balance;
- quantities of material to be moved;
- direction of movement areas where earth may be borrowed or wasted;
- quantities involved; and
- best policy to adopt to obtain the most economic use of plant.

The MHD is a curve that we plot on a distance base. The ordinate from the base line at any point represents the algebraic sum of the volumes of cuts and fills, from the start of the earthworks to that specific position.

 This explanation will become more meaningful when we work through an example.

The sign convention used is:
Cuts + ve
Fills – ve

See Fig 11.1 on page 205 for clarification.

11.3.2 Preparation

Before we can draw the MHD, we calculate the volumes. We follow the steps below before drawing the MHD.
1. A decision is made with regard to traffic volumes and design speed, to determine what design parameters are to be adopted.
2. From the design parameters, the maximum gradients are adopted.
3. Possible routes are selected, while taking into consideration the above restrictions as well as the cost of construction.
4. A longitudinal section is drawn.

The road formation line (which is equal to the final level or formation level of the road) is then plotted onto the long section with due regard to the design parameters. This will indicate where the positions of cut and fill are (see Fig 11.1 on page 205).

Cross-sections are now drawn. Cross-sections are drawn perpendicular to the longitudinal section on straights, and radially on curves. These cross-sections supply information on the slope of the ground (cross fall) on either side of the longitudinal section, and supply data for the calculation of earthworks quantities. To calculate these volumes you need to review your survey and maths knowledge. We plot the co-ordinates of the sides of the sub-grade and the tops and bottoms of embankments at predetermined chainages, and draw the cross-section and calculate the cross-sectional areas. We then calculate the

Module 3: Unit 11

approximate volume of earthworks. For this, we use the **end area method**, as described on page 180 at the start of this module. We now have a list of volumes to be excavated or tipped at the predetermined chainage all along the route. From this, we can draw the MHD.

11.3.3 Bulking and shrinkage

Bulking

Excavated volume will be greater than its 'in situ' volume. Depending on the material, we might have an increase in bulk from 10% (in the case of sand) to 50% (in the case of rock).

Shrinkage

If compacted, the same material may occupy less volume than when originally 'in situ'. To take this into account, we have to apply a correction factor to cut and fill volumes (see Example 11.1 on page 207).

11.3.4 Properties of the MHD

Consider Fig 11.1 below. The ground XYZ is to be levelled to a grade line A'B'. Assuming that the fill volumes, after correction, equal the cut volumes, the MHD would plot as follows:

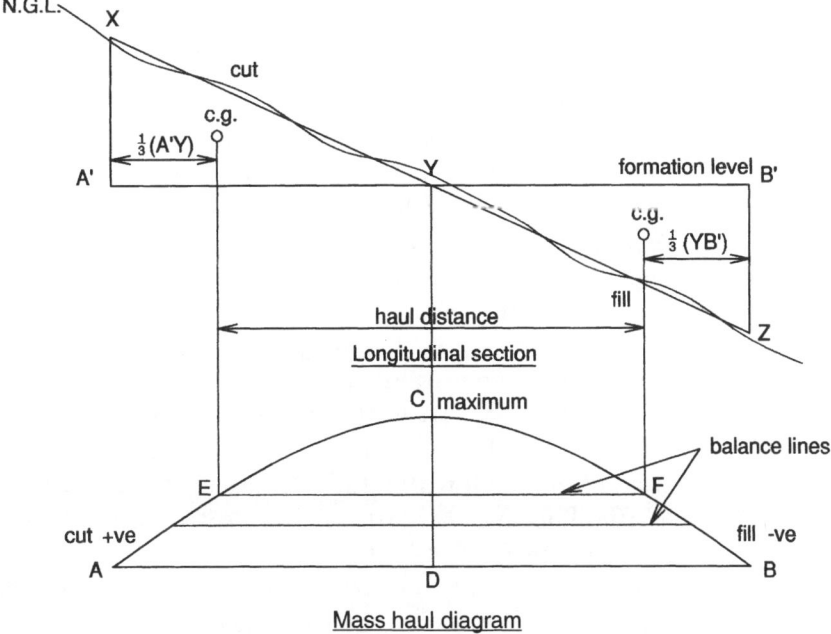

Fig 11.1 Properties of MHD

205

The properties of the MHD will become clear when we work through Example 11.1.

Since the curve of the MHD represents the algebraic sums of the volumes, any horizontal line drawn parallel to the base AB will indicate the volumes that balance. Such a line is termed a 'balance line', which indicates that the total cut equals the total fill.

The curve rising from the left to the right indicates 'cut', while the curve falling indicates 'fill', hence cut (+ve represents positive) and fill (−ve represents negative). The maximum and minimum points of the MHD occur directly beneath the intersection of the natural ground and the formation grade line. These intersections are termed 'grade points'. The maximum ordinate CD represents the total cut volume.

 Note that in our MHD there are no minimum points.

The **haul distance** is measured from the centroid of the cut volume to the centroid of the fill volume. The total haul in the section is the total volume (CD) moved through the total distance (EF).

11.3.5 Balancing procedure

To illustrate the use of **freehaul** distance, consider the following with reference to Fig 11.2: X (see page 207).
1. Assume that the contractor specifies a freehaul distance of 100 m. Use a scale ruler to move this distance parallel to the base line until it cuts the curve at E and F.
2. EF represents the position on the longitudinal section where cut LMY equals fill volume YNP (the quantity equal to CC).
3. The remaining cut volume XLMA' is represented by the ordinate EG, and is the **overhaul** volume.
4. The overhaul volume XLMA' has now to be filled into ZPNB', the average distance being from centroid to centroid. The positions of these centroids are found by bisecting EG and FH, to give the horizontal distance JK between centroids.
5. Assuming JK = 250 m, the overhaul volume has to be moved through this distance. However, as mentioned previously, the first 100 m is still within the freehaul contract specified, thus the overhaul distance is (250 − 100) = 150 m. It is obvious that the total volume (CC + EG) = CD.
6. Thus, the overhaul = overhaul volume moved through the overhaul distance.
   = EG moved through (JK − EF)

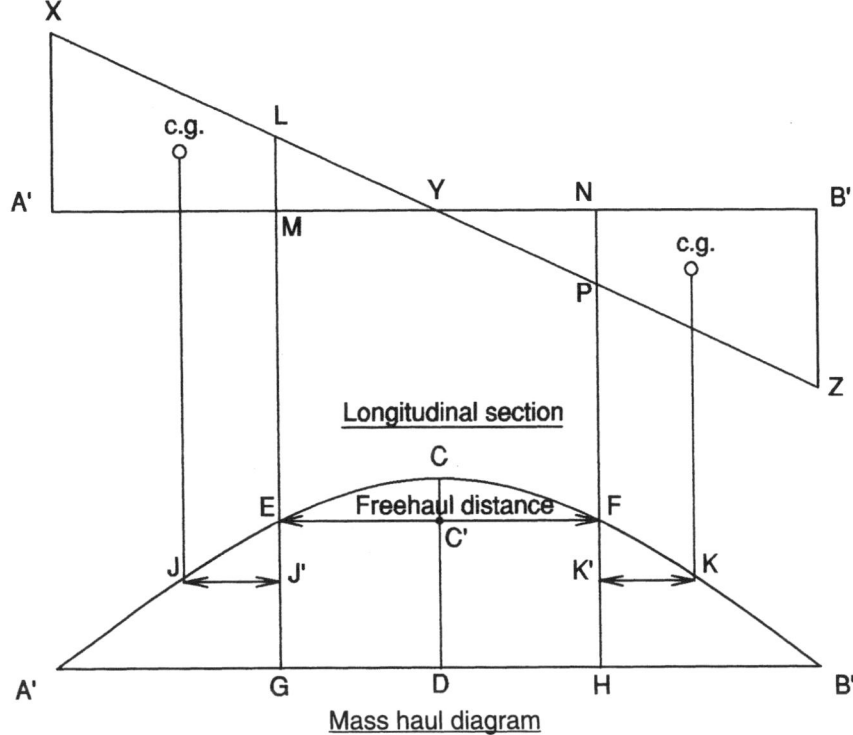

Fig 11.2 Freehaul distance

11.4 Summary

This section deals with the planning of projects such as roads, dam walls, pipelines, and the like. This section also deals with the positioning of these structures on a contour plan. To build these structures you will have to determine which position will be the most economical. You also need to be able to determine the estimated cost of the proposed project. This means that you will have to know the amount of soil that will be needed, either to fill or to excavate and the plant needed, which means that you must be able to calculate the quantities, even before you physically start with the project.

SOLVED EXAMPLES

EXAMPLE 11.1

You are required to do the planning of earthworks on a road contract over a section of 350 m, disregarding the sections on either end. Volume at station 70 is zero.
- The given table shows the stations and existing ground levels along the centre line of the road.

- The formation level at station 70 is 43,5 m above datum, and rises uniformly at a gradient of 1,2%.
- The volumes are recorded in cubic metres, cut being (+ve) and fills (–ve).
- The stations are situated 50 m apart.

1. Plot the longitudinal section using a horizontal scale of 1:2 500 and a vertical scale of 1:250.
2. Assuming a correction factor of 0,8 applies to fills, plot the mass haul diagram to a vertical scale of 1:50 and a horizontal scale of 1:2 500.
3. Calculate the total haul, and indicate the haul limits on the MHD.
4. Assume a freehaul of 300 m.
   a. What is the freehaul volume?
   b. What is the overhaul distance?
   c. Calculate the overhaul.

Station	Ground level	Volume
70	52,8	
		+1 860
71	57,3	
		+1 525
72	53,4	
		+547
73	47,1	
		–238
74	44,7	
		–1 000
75	39,7	
		–2 025
76	37,5	
		–1 572
77	41,5	

Solution: Longitudinal section
See Fig 11.3 on page 210.

From the longitudinal section, you can see that cut changes to fill halfway between stations 73 and 74, or 175 m from station 70.

Module 3: Unit 11

Station	Ground level	Volume	Corrected volume	Mass ordinate
70	52,8			0
		+1 860	+1 860,0	
71	57,3			1 860,0
		+1 525	+1 525,0	
72	53,4			3 385,0
		+547	+547,0	
73	47,1			3 932,0
Change from cut to fill		−238	−190,4	
74	44,7			3 741,6
		−1 080	−864,0	
75	39,7			2 877,6
		−2 025	−1 620,0	
76	37,5			1 257,6
		−1 572	−1 257,6	
77	41,5			0

Remember that you have to apply a correction factor to fills. In this question, you will have to apply this factor halfway between stations 73 and 74. Thus, the corrected volume = −238 × 0,8 = −190,4 m³

Remember to do the same as done in the table above at all the other fill volumes. The mass ordinate is simply the cumulative volume. Now, you should be able to plot the MHD. See Fig 11.4 on page 211.

The dotted line on the MHD is the total haul distance = 322,5 m and the total volume, CD = 3 950 m³.

Therefore, total haul = 3 950 m³ moved through a distance of 322,5 m.
1. Freehaul volume = 3 340 m³ (scaled off/read off MHD)
2. Overhaul distance = Total haul distance − freehaul distance
   = 322,5 − 300
   = 22,5 m
3. Overhaul volume = 3 950 − 3 340 = 610 m³ (= ED)

Therefore, the overhaul = 610 m³ moved over 22,5 m.

 As you can see, mass balance calculations are a lot of work and quite complex. It is no surprise that practising engineers today use computers to do these complicated calculations for them.

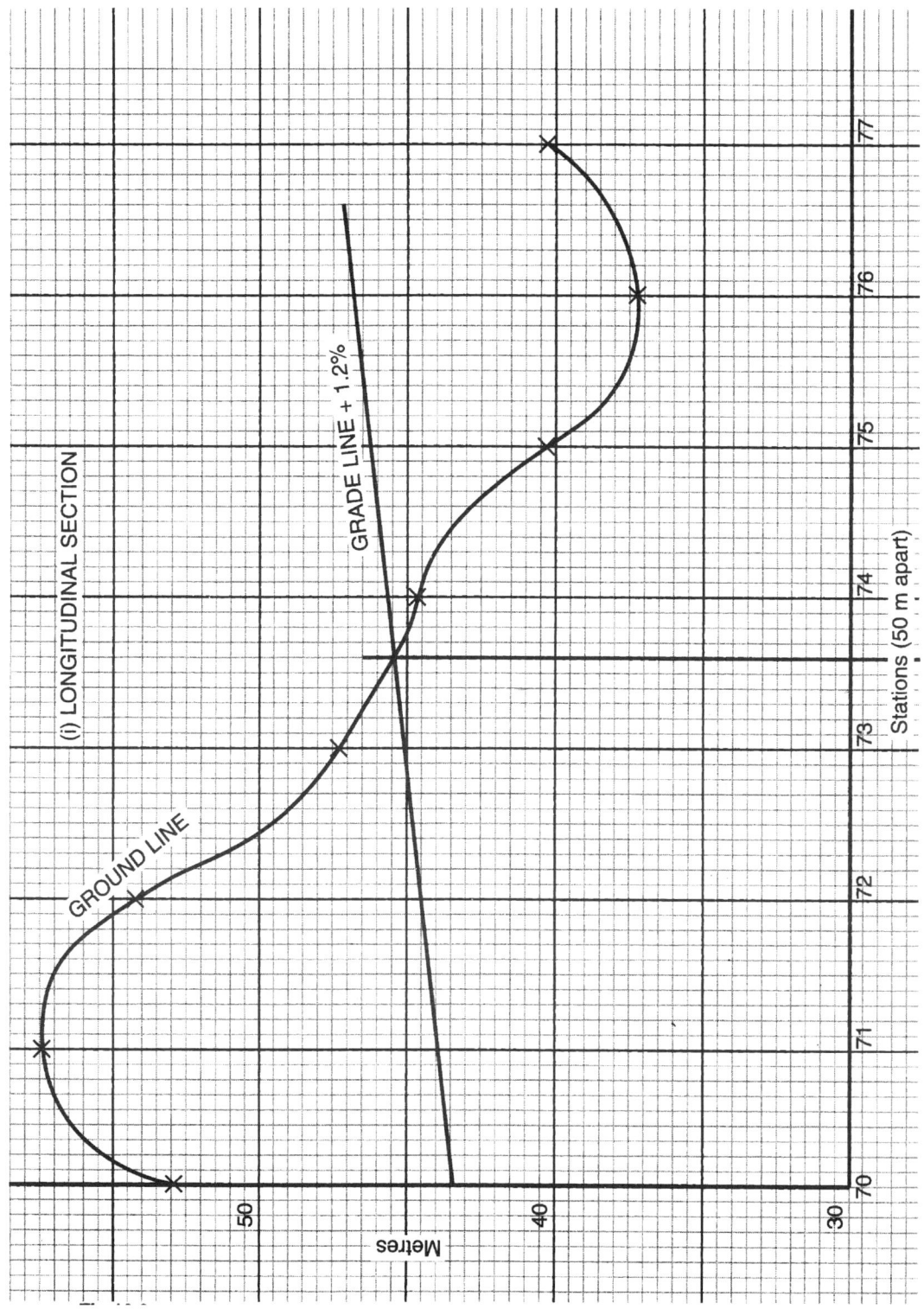

Fig 11.3 Longitudinal section

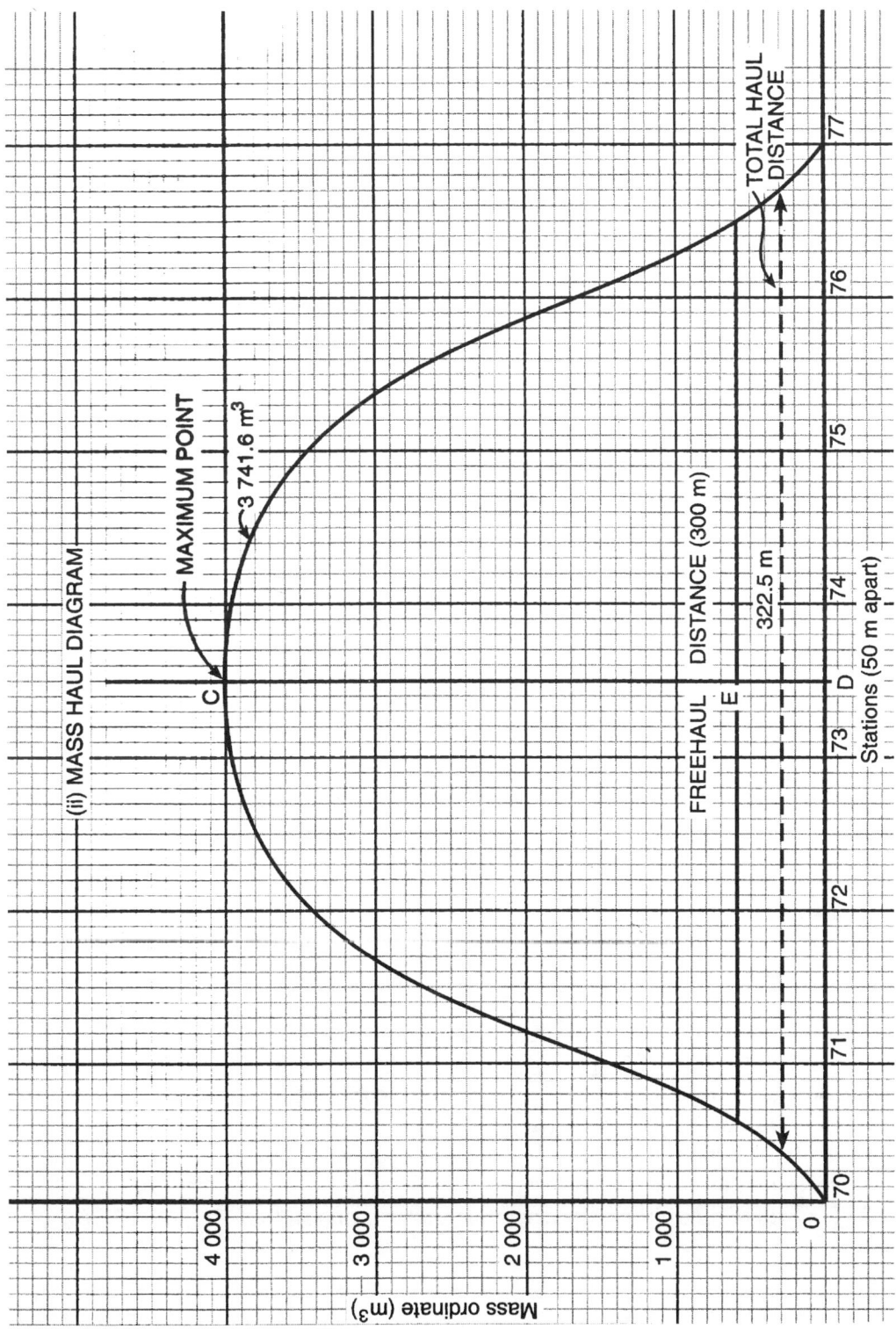

Fig 11.4 Mass haul diagram

Drawing for Civil Engineering

Let us clarify what we have discussed thus far by continuing with the example from the previous unit.

EXAMPLE 11.2

Part B: The longitudinal section
(See Part A: The contour plan, from Example 10.1 on page 199.)
You have to design a sewer pipeline to run from point C to a bend at point D and from there to the corner E of a sewer pump station, denoted by corners E, F, G and H. Manholes are to be shown at points C and D. See Fig 11.9 in the pocket at the back of the book.

The co-ordinates of these points are as follows:

Point	Y	X
Constant	−80 000,00	+200 000,00
D	−1 300,00	+4 550,00
E	−1 180,00	+4 550,00
F	−1 150,00	+4 550,00
G	−1 150,00	+4 525,00
H	−1 180,00	+4 525,00

1. The invert level of the pipe of the 160 diameter µPVC sewer pipe at C = 6,88 m and at D = 5,77 m. The grade (slope) of the sewer pipe from D to E is 1:110. The level at the bottom of the pump station is 2,25 m.

 Who knows the meaning of the abbreviation µPVC?

- Draw a longitudinal section, horizontal scale 1:1 000 and vertical scale 1:100, for the sewer line. C-D-E showing chainage, ground level, invert level, depth, grade/pipe details, area and volume, at 20 m intervals.
- Point C is at chainage 0,00 m.
- Neatly print the title and scale below the drawing.
- The width of the sewer trench is 1,2 m.
2. Calculate the volume of the excavation in cubic metres (m³), using the end area method.
3. Calculate the volume of excavation for the pump station.

Solution

1. a. We obtain the lengths of pipes CD and DE by calculating the joins between the respective points. See Fig. 11.9 in the pocket at the back of the book.
   b. Let us work through one area and volume calculation.

   Area calculation at manhole C
   = depth of trench × width of trench
   = 1,70 × 1,2 = 2,04 m³

   Depth of trench = ground level – invert level

   Volume between chainage 0 m and 20 m
   = average area × distance
   = $\frac{(2,04 + 2,08)}{2}$ × 20 mm
   = 41,20 m³

   Continue at 20 m intervals until you get to manhole E.

2. Volume of the trench excavation
   = sum of the volumes between manhole C and pump station
   = 528,41 m³

3. Volume of pump station excavation:

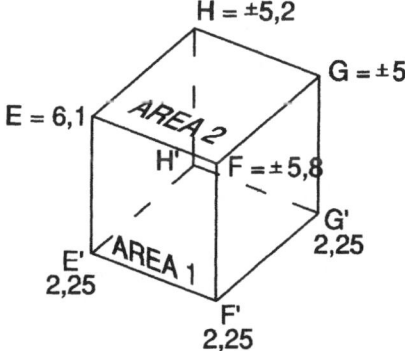

Fig 11.5 Volume calculation diagram

a. The ground levels have been interpolated to find the values as indicated on the sketch for E, F, G and H.
b. To calculate the plan dimensions of the pump station, do a join calculation from E to F and E to H.
c. The plan dimensions of the pump station are 25 m × 30 m. You have to calculate the average areas of the sides, as follows:

Height EE' = 6,1 − 2,25 = 3,85 m
Height FF' = 5,8 − 2,25 = 3,55 m → Average height = (3,85 + 3,55)/2
= 3,70 m

Height GG' = 5,0 − 2.25 = 2,75 m
Height HH' = 5,2 − 2,25 = 2,95 m → Average height = (2,75 + 2,95)/2
= 2,85 m

Therefore:
Average area (1) = 3,70 × 30 = 111,0 m²
And average area (2) = 2,85 × 30 = 85,5 m²

Volume of pump station excavation = $\frac{(111,0 + 85,5)}{2} \times 25$

= 2 456,25 m³

EXAMPLE 11.3

The contour plan in Fig 11.10 in the pocket at the back of the book shows the centre line of a portion of a proposed road which is 10 m wide. This contract starts at point A, which is at stake value 1 000 m and ends at stake value 1 200 m.

The grade of the road is constant between stake value 1 000 m and the beginning of the vertical curve (BVC 1), between end vertical curve (EVC 1) and BVC 2, and between EVC 2 and stake value 1 200 m.

Module 3: Unit 11

Stake value	Ground level	Road level	Remarks
1 000	231,000	228,200	Start of contract
1 020	228,67	Interpolate	Constant slope
1 025		227,688	BVC 1
1 040	225,75	227,675	
1 055			PI = 227,074
1 060	223,25	228,575	
1 080	228,21	230,523	
1 085		231,174	EVC 1
1 100	232,07	Interpolate	Constant slope
1 115		235,273	BVC 2
1 120	238,30	235,925	
1 140	243,38	237,905	
1 150			PI 2 = 240,057
1 160	242,46	238,880	
1 180	239,79	238,851	
1 185		238,687	EVC 2
1 200	235,07	238,100	End of contract

1. Draw a longitudinal section of the road between stake value 1 000 m and 1 200 m.
   - Use a horizontal scale of 1:500 and a vertical scale of 1:200.
   - Give the following data at 20 m intervals: stake value, ground level, road level, the grade of the road, cut/fill depths and the curve details, that is, the stake values, levels and so forth of all BVCs, EVCs and PIs.
   - Curve details should have been done in *Surveying*.

Do not forget to calculate the crest and sag positions, and values of the vertical curves.

You will notice that in this example, the ground levels have been given, but these values should be interpolated by using the contour plan.

2. Make a neat sketch of a mass haul diagram for this section of road, below the longitudinal section.
3. Draw cross-sections of the road at stake value 1 120 m and stake value 1 140 m.
   - Determine the approximate volume of cut material to be excavated between these two stake values.
   - Show area and volume calculations.
   - Use a scale of 1:200.
   - The side slopes for cut and fill is 1:2; assume the road is flat in cross-section.

4. Draw the plan of the road and its embankments on the contour plan for the section of road between SV 1 000 m and SV 1 100 m only.

You will need the assistance of your lecturer to fully understand the method used for completing this exercise.

As an additional exercise, you could complete the embankment between SV 1 100 m and SV 1 200 m. Fig 11.6 may be used to determine the shape of the embankment in plan.

Method to indicate an embankment when in:
a. fill

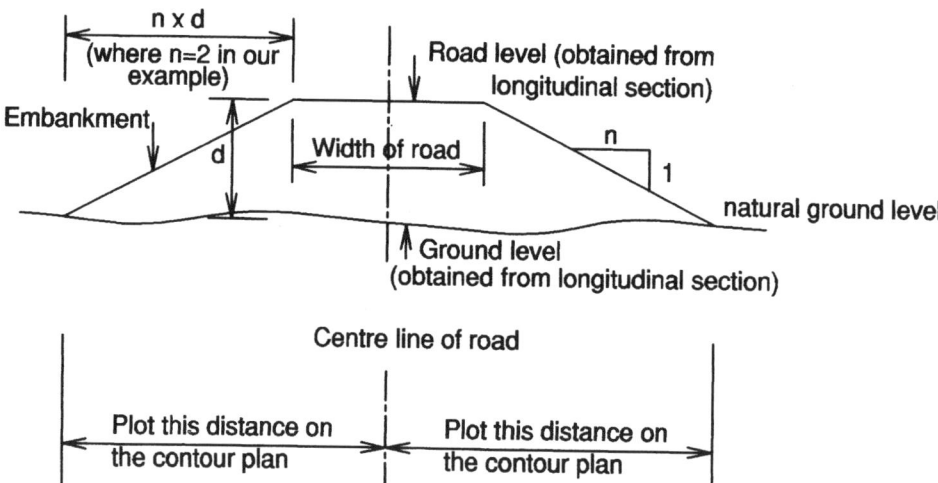

Fig 11.6 Cross-section at any chainage

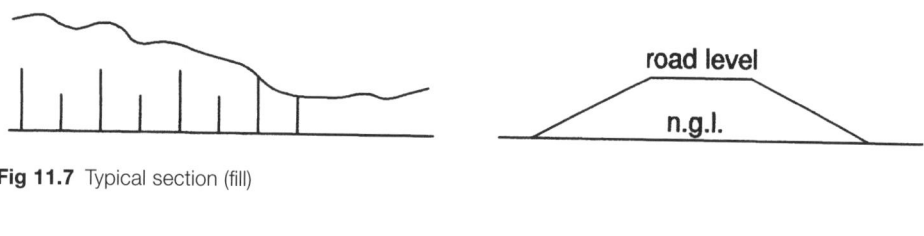

Fig 11.7 Typical section (fill)

b. cut

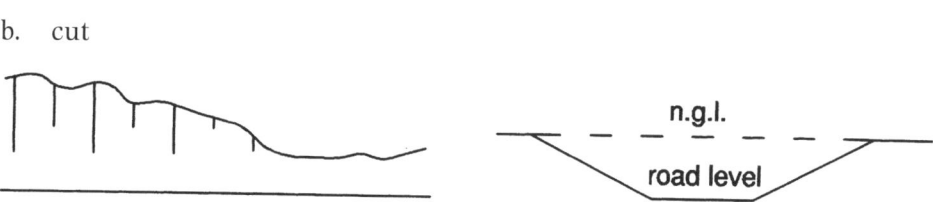

Fig 11.8 Typical section (cut)

216

Module 3: Unit 11

Please refer to the following figures, which can be found in the pocket at the back of the book, for the solutions to the four parts of this example.
- Fig 11.12;
- Fig 11.13;
- Fig 11.14;
- Fig 11.15; and
- Fig 11.16.

Activity 11.1

(Answer for (2) included)

You will need to have completed the contour plan in Activity 10.1 of Unit 10 to continue with this activity. You have to design a proposed road, which is 10 m wide, to run in a straight line from survey station A to survey station C.
- Both these points fall on the centre line of the proposed road.
- Station A is at stake value 0,000 m.
- The slope (grade) of the road is constant between station A and station C.
- Plot the centre line of the proposed road on the contour plan.
- The road level at point A = 33,018 and the road level at point C = 32,061.

1. Draw the longitudinal section of the road between Station A and Station C. Use a horizontal scale of 1:1000 and a vertical scale of 1:100.
   - Give the following data at 20 m intervals: stake value (distance), ground level, road level, cut and fill depths, the slope of the road, the area and the volume.
   - Neatly print the title and scale below the drawing.

2. Calculate the volume of cut and fill material in cubic metres, using the end area method.
   - For the purpose of cross-sections, assume that the ground level and road level over the whole cross-section is the same as the ground level and road level at the centre line of the road.
   - The slopes of the embankments in both cut and fill can be taken as 1:1.
   (Cut = ± 3 406 m³; Fill = ± 2 422 m³)

217

Activity 11.2

(Answers for (3) included)

Fig 11.17, which can be found in the pocket at the back of the book, shows the contour plan of the centre line of a proposed dam wall, which is to be built from A to B.
- As shown, the scale of the contour plan is 1:500 and the contours are at 2 m intervals.
- The level at the top of the wall is 128 m and the width of the wall at the top is 5 m.
- The slopes of the upstream and downstream embankments are given on the contour plan.

1. Draw a plan view of the wall on the contour plan provided.
2. a. Draw a longitudinal section of the wall (along the centre line shown) to a horizontal scale of 1:250 and a vertical scale of 1:100. Take point A as chainage zero.
   b. Provide a table under this longitudinal drawing, giving the following information at 10 meter intervals: chainage, ground level, height of fill, cross-sectional areas, and volume.
3. Calculate the volume of fill material required to construct the wall by using the end area method (± 8 830,37 m³).

Self-evaluation

1. Which specifications are you going to use when designing structures such as roads, railways and the like?
2. How do you differentiate between cut and fill, and why are they necessary?
3. Discuss the purpose of a mass haul diagram.
4. Define:
   a. longitudinal sections; and
   b. cross-sections.

SELF-EVALUATION ANSWERS

1. The specifications issued by the Department of Transport and the Provincial Administration Standards.
2. Roads and such structures should be built to predetermined grades to enable smooth traffic flow but, unfortunately, the natural ground level is not often suitable. Consequently, soil sometimes needs to be removed (cut) from one position to another, and/or while soil sometimes has to be used to fill hollows.
3. The mass haul diagram is a diagram that is mainly used to determine where you either have excessive soil, or where you need soil after you have designed your line structure. This diagram indicates the accumulative volume of soil at any point along the centre line of the structure.
4. a. Longitudinal sections are sections that run along the centre line of a construction project.
   b. Cross-sections are sections perpendicular to the centre line of a construction project.

Unit 12 Introduction to CAD in surveying

For this unit you will need the assistance of a lecturer or somebody acquainted with this type of software.

12.1 Introduction to SURPAC SURVEY SOFTWARE

This program provides Computer Aided Drawing facilities using the current Co-ordinate File as the source for Co-ordinated Points.

12.1.1 Co-ordinate File

SURPAC SURVEY SOFTWARE, which is the software package that will be used in this course, requires the creation of a Co-ordinate File for each project. Co-ordinate Files contain Co-ordinate information, or Elements, of the following type:

- Point Name (1 to 8 characters);
- Y Co-ordinate;
- X Co-ordinate;
- Height, or Elevation;
- Auxiliary Height;
- Point Description — (1 to 8 characters, or a number 0 to 499) (see below);
- Field Book Reference;
- Calculation Page Reference; and
- Date.

Not all the above Elements have to be displayed. Elements may be selected to suit the requirements of any given Co-ordinate File. SURPAC must have a Co-ordinate File loaded at all times. This File is known as the Current Co-ordinate File. All Co-ordinate File-related operations will relate to the Current Co-ordinate File. Co-ordinate Files may be specified as being in either the Engineering or Cadastral format. For the Engineering format, the Point Description is either a code number from 0 to 499, or an alphanumeric entry of up to 8 characters. Refer to the Beacon Description List. For the Cadastral format, the Point Description is a code number from 1 to 55, followed by a single alphanumeric character. Refer to the Beacon Description List. The maximum size of a Co-ordinate File is 32 600 Points. Co-ordinate Files have the extension .wcr.

12.1.2 Opening an existing or creating a new Co-ordinate File

On loading SURPAC, the first display normally shown is a File Display Window that displays the contents of the current or defined

Co-ordinate File. If a new Co-ordinate File is defined then the Program will prompt you for the entry of the File's System. In this course, we will use co-ordinates on or near the MUT campus, on System WG31. Various other items of information are also entered for a new File, and each entry is self-explanatory.

12.1.3 Displaying the Co-ordinate File

The File Display Window is variable, with the display columns being set via the Co-ordinate File Elements item on the File Main Menu heading. The Point Name, Y co-ordinate and the X co-ordinate are compulsory display items, and the columns associated with them will always be displayed. The other items making up the Co-ordinate File are optional.

The information for any of the Points displayed in the File Display Window may be Edited by clicking the appropriate Line in the grid box, and then using the Edit Window. The Points currently contained in the Co-ordinate File are displayed in the alphanumeric order of the Point Name. This alphanumeric sequencing is automatically maintained throughout the SURPAC session, regardless of which new Points are Added or Deleted; and regardless of the order in which the Points are actually stored in the disk File. This alphanumeric listing makes it easier for the user to visually locate a required Point but, more importantly, it ensures that the Program is able to rapidly locate any Point regardless of the File size (the File size should not exceed 32 600 Points). The Name of the currently loaded Co-ordinate File is displayed in the first panel of the SURPAC Status Bar. Clicking this item in the Status Bar enables you to change to a different Co-ordinate File.

12.1.4 Auto-Compilation of an Engineering Format Beacon Description List

This function is only applicable to Engineering Format Co-ordinate File Beacon Description Lists. The purpose of this function is to allow for an automatic re-compilation of a Beacon Description List for a Co-ordinate File for which the Descriptions used in the Co-ordinate File do not match those held in the Beacon Description List. If a Co-ordinate File is imported from an ASCII File, for example, the Descriptions made by the Points in the imported File may bear no resemblance to those held in the Default Beacon Description List. The Default Beacon Description List is the List of Beacon Descriptions that is automatically attached to a new Co-ordinate File when that File is first created. To Auto-Compile the Beacon Description List for a Co-ordinate File, do the following:
1. While in the 'Co-ordinate File Editing' Program, click the 'Descriptions' menu item in the 'Actions' Bar Menu.
2. If all the Beacon Descriptions used by the Points in the Co-ordinate

File are included in the Beacon Description List, then the Beacon Description List form will be displayed in the usual manner.
3. If not, a prompt will ask you if you want SURPAC to recompile the Beacon Description List in compliance with the Descriptions that are used by the Points in the Co-ordinate File. If you reply 'Yes' to this prompt, then a new Beacon Description List will be generated, which is derived from the Point Beacon Descriptions already existing in the Co-ordinate File. This process is advised before using the General CAD Program.

Activity 12.1

Create the following Co-Ordinate File:
Co-ordinate File = MUT2009 (or assign a name of your choice)

System: WG31

Name	Y Co-ordinate	X Co-ordinate	Height	Description
ALDEN	8 244,452	3 316 792,497	99,076	TRIG
C1	8 521,670	3 317 003,631	78,398	PEG
C2	8 524,381	3 316 937,330	81,316	PEG
C3	8 458,293	3 316 861,796	81,438	PEG
C4	8 413,697	3 316 956,271	78,355	PEG
C5	8 424,479	3 317 054,548	77,414	PEG
C6	8 372,510	3 316 949,007	90,391	PEG
COOP-R	−504,230	3 312 835,470	0,000	RES
M1	8 383,657	3 316 926,939	91,663	PEG
M2	8 386,707	3 316 915,101	92,324	PEG
M3	8 389,541	3 316 893,745	93,955	PEG
TR125	10 638,050	3 316 486,870	174,600	UMGODWE
TR220	−485,300	3 312 982,470	136,900	CO-LT
TR297A	8 380,146	3 316 887,451	99,019	BCN
TR343	7 396,340	3 321 665,460	133,100	ATHLONE
TR347	14 578,130	3 332 819,120	82,200	ILL RES
TR349	11 142,430	3 324 165,260	108,300	AMRES
TR350	12 784,580	3 327 826,060	128,300	ELGRO
TR351	11 769,640	3 322 955,730	137,000	GARDEN
TR352	14 600,540	3 323 654,440	183,400	KWA MAKH
TR524	4 737,520	3 311 598,100	142,700	WOODLAND

Name	Y Co-ordinate	X Co-ordinate	Height	Description
TR547	3 238,910	3 317 999,110	107,000	ISIP-NEW
TR560	6 749,810	3 315 575,800	102,500	UMLAZI
TR587	2 407,590	3 316 811,870	89,500	MOR-AUX
WT1	8 129,730	3 318 070,210	0,000	RES17
WT2	7 004,110	3 316 609,510	0,000	RES18
WWT	4 712,660	3 311 548,930	0,000	TOWER

Your lecturer will assist you by guiding you through the steps, as described in the pages above. Note that when you set up the Co-ordinate File parameters, you must decide whether it is advisable to use co-ordinate constants for either your Y and/or X values.

12.1.5 Printing the Co-ordinate File

VERY IMPORTANT: From the General Menu, select 'Print the Co-ordinate File List'. After selecting relevant options, click OK. From the next window, select 'ASCII FILE'. Save the coordinate File in ASCII format '.lst'.

12.1.6 Selecting the 'Set Text File (*.vpg) as Printer' option from the File Menu

This selection sets the Virtual Printer mode and allows you to 'print', or store, the Calculation pages onto a disk ASCII File. The default page size is the page size (such as A4) that was previously set when selecting a Printer/Plotter. The Name of the ASCII File is the Name of the current Co-ordinate File plus the extension '.vpg'.

VERY IMPORTANT: Select this option before doing Polar and Join Calculations.

Calculating a polar
1. Program Description
   The Horizontal Direction and Reduced Distance are used to compute the [Y, X] Co-ordinates of the new Point.
2. Program operation
   The Program presumes that the Data to be entered consist of a Setup Point Name, which may be selected from the Point Name Display Window; an oriented horizontal direction; and a reduced Distance. The Program will prompt for the entry of these terms into the appropriate Input Boxes. The Program will prompt for the entry of the new Point's Description, but does not exclude other items from being entered that may be required.

All Data are entered via the keyboard. The keyboard is the only mode of Data entry when using the Plane Data Input mode. Once all Data have been entered, enter the Name of the Point to be intersected and enter, or verify, the Point Description.

When the AutoPrint Button is clicked, the AutoPrint mode will be set to ON, and will remain ON until this button is again clicked. All Polar Calculations done while the AutoPrint mode is ON will automatically be sent to the Current Printer. When the AutoPrint mode is ON, the Print Polar option becomes disabled.

Clicking the Print Polar Button (when printing is enabled) will cause the currently displayed Calculation to be sent to the Current Printer.

Calculating a join
1. Program description
   The Horizontal Direction and Horizontal Distance are computed from the two pairs of [Y, X] Co-ordinates.

   The Names of all the Points in the Current Co-ordinate File are displayed in a Point Name Display Window, located at the bottom of the screen. Clicking a Point Name in this window will allocate that Point Name to the Input Box that is waiting for Data entry (that is, either the 'From' or 'To' Point), and the Co-ordinates of the Point, Height and Description to the appropriate Display Boxes.

   Point Names may be entered from the keyboard, if required. First, ensure that the flashing text cursor is in the appropriate Name Input Box before you type in the required Name.

   Once both the 'From' and the 'To' Points have been entered, the Program will display the results of the Calculation in the appropriate Calculation Display Boxes.

   The Options Button allows you to set the angular display to the nearest 1' (nearest one minute), or to the nearest 0.1' (nearest one minute).

   When the AutoPrint Button is clicked, the AutoPrint mode will be set to ON, and will remain ON until this button is clicked again. All Join Calculations that are done while the AutoPrint mode is ON will automatically be sent to the Current. When the AutoPrint mode is ON, the Print Join option becomes disabled.

Activity 12.2

Import the Co-ordinate File and Calculations into Ms Word:
1. Open MsWord 2007.
2. Select 'Insert' from the ribbon.

3. Click on the arrow next to 'Object'.
4. Select 'Text from File'.
5. Change 'Files of Type' at the bottom of this window to 'All Files'.
6. Select your own .lst File and double click.
7. Select your own .vpg File and double click.
8. Format the MsWord document.
9. Write a short report on this project, and submit by e-mail.

12.2 The General CAD Construction/Edit/Plotting Program

This Program provides Computer Aided Drawing (CAD) facilities using the current Co-ordinate File as the source for Co-ordinated Points. A CAD Plotting Sheet is created using User-defined parameters.

On the CAD Plotting Sheet, drawing facilities such as Line and Arc Construction and Editing, Point Naming using various Symbol Marking, writing and Editing Text Items, drawing Grids, Hatching Polygons, plotting Spot Heights, and the automatic positioning and writing of Line Directions and Distances are available.

Using the current Co-ordinate File as the source for Co-ordinated Points, the Program provides CAD facilities. A CAD Plotting Sheet is created using User-defined parameters.

The Program is run by selecting the appropriate Menu Item under the General Main Menu Heading. Initially, the Program will load, or Setup, the CAD Plotting Sheet, having the same Name as the current Co-ordinate File.

To Setup a CAD Plotting Sheet, select the Plot File Sheet Setup Information Menu item, found under the File Main Menu Heading.

It is necessary to Setup a CAD Plotting Sheet by defining certain parameters of the sheet. These variable parameters consist of the Y and X Co-ordinates of the Sheet Centre, the Size (Length and Height) of the sheet, the sheet Swing, the sheet plotting Scale and the sheet Margins. The Setup Program will determine default values for the above parameters, as follows.

The default Sheet Centre will be calculated at the centroid of the Points existing on the Co-ordinate File, excluding any Trigonometrical Beacons or Town Survey Marks. The plotting scale will be computed to fit the entire File on a sheet of $1189\,mm \times 840\,mm$, less the margins. Therefore, if you want to make use of this default routine, make sure that the Co-ordinate File does not contain extraneous Points (such as far-away terminals) that will affect the determination of the Sheet Centre and the sheet scale. Obviously, any values setup via this default routine may be Modified via the Setup Form Input Boxes.

Use the appropriate Input Boxes to enter or Modify any of the sheet parameters. By clicking the Printer Plotter Font Name Display Box, you can select any of the available True Scale Fonts. As a default Font, SURPAC will use the Arial True Type Font for the CAD Sheets.

12.3 CAD Plotting Sheet Construction/Editing

The main Construction and Editing functions carried out in this Program are:
- Line and Arc Drawing;
- General Text Writing;
- Line Direction and Distance Data Writing;
- Marking Points with defined Symbols and Displaying their Names;
- Hatching of Polygons;
- Image Adding and Editing;
- Grid Intersection Marking and North Point Drawing; and
- Creating and/or Editing Point Data from Line Information.

In this course, we will only consider Line and Arc Drawing, and General Text Writing.

12.3.1 Line and Arc Drawing

Lines are drawn by using a combination of cursor movement together with the Start — [S] and End — [E] Commands.

Start Command
The function of this Command is to determine the Start position for the construction of a Line, an Arc, a Rectangle and the like. The Command is activated either by clicking the Mouse [LHB] or by pressing the 'S' key on the Keyboard (using 'CAPS ON' mode).

End Command
The function of this Command is to determine the End position for the construction of a Line, an Arc, a Rectangle and the like. The Command is activated either by clicking the Mouse [RHB] or by pressing the 'E' key on the keyboard (using 'CAPS ON' mode).

Snap Radius
You may set the Snap Radius to any value between 1 and 15 mm. This determines how far the Program must search for Data from the current cursor position. Now, when the Start — [S] command is given, the Program will search for any Co-ordinate Points (as held on the current Co-ordinate File) that are within range of the Snap Radius. If no Points are found, the Program will search for any Line terminals that exist within the Snap Radius.

The cursor will then snap onto the closest of any Points (or Line terminal) found, and the new Line will be drawn from this position. If no Point (or Line terminal) is found, the new Line will be drawn from the screen position defined by the current cursor position. As the cursor is moved (either by the Mouse, or by the Keyboard movement Keys) a construction Line will be produced from the Start Point, which follows the cursor position.

Once the cursor has been moved to the required position, give the End — [E] Command. The Program will again search for any fixed Points within the Snap radius of the cursor position. If no Points are found, the Program searches for Line terminals. If a Point (or Line terminal) is found, it will snap onto the closest one. The new Line will then be drawn using the current Line Colour (or Pen number), Style and Thickness.

Arcs, or circular Curves, are drawn in a similar manner, with the inclusion of an intermediate Point that indicates any position on the circumference of the Arc. Use the Start Command, as described above for Lines. Then move the cursor to a screen position through which you would like the arc circumference to pass. At this Point click either the Mouse [MB] (middle button) or the [\] on the Keyboard. Now, when the cursor is moved, the screen will display a 'rubber band' Arc that starts at the defined Start position, passes through the defined circumference position and ends at the current cursor position. Move the cursor to the required position for the End of the Arc, and then give the End Command, as described above for Lines. The new Arc will then be drawn using the current Line Colour (or Pen number), Style and Thickness.

SURPAC accommodates five different Line Styles. These are as follows:
1. SOLID LINE
2. DOTTED LINE
3. DASHED LINE
4. DASH-DOT LINE
5. DASH DOT-DOT LINE

The Line Style, Line Colour and Line Thickness options are accessed either by clicking the appropriate Menu Item, or from the fixed Toolbars.

Lines or Arcs are Erased (or Rubbed) by placing the cursor close to the required Line or Arc, and by then giving the Rub Line [RL] Command or the Rub Arc [RA] Command.

12.3.2 General Text Writing

Text Items of up to 80 characters in Length may be written anywhere on the current Sheet. Each Text Item has user-definable attributes, which are:
- the Text Direction (in Degrees, using the Standard Survey Convention);
- the Text Size (in mm);
- the Text Colour;
- the Text Width To Height Ratio (as a number between 0,1 and 1,0);
- the Text Mode, which is Normal, Italic, Bold or Bold Italic; and
- the Underline mode (Underline or normal).

Text Items may be copied using the Text Copy [WC] Command. While a Text Item is being copied, it may also be rescaled by using the Increase Text Size [+] Command, or the Decrease Text Size [−] Command, or it may be re-oriented using the Increase Text Direction [>] or Decrease Text Direction [<] Commands.

Text Items may be moved using the Text Move [WM] Command. While a Text Item is being moved, it may also be re-scaled by using the Increase Text Size [+] Command, or the Decrease Text Size [−] Command, or it may be reoriented using the Increase Text Direction [>] or Decrease Text Direction [<] Commands.

Text Items may be edited by using the Text Edit — [WE] Command. The Text Item itself may be edited, as may any or all of the attributes of Text Items.

A mass Text Item Editing facility exists for changing all Text Items, or their attributes. This Text Item Editing facility uses the Reset Text Attributes [WR] Command to implement changes. A variation of this command, namely the Reset Frame Text Attributes [BT] Command, allows a framed or limited screen area to be used for Text Editing.

A Text Item may be Erased (Rubbed) by using the Rub Text Item [RW] Command. Text Items that have been Erased, or Rubbed, during the current session may be retrieved and redisplayed by using the Undelete Text Item [WZ] Command. Text Items are retrieved in the reverse order in which they were erased. So, the last Text Item that was removed will be the first Text Item to be retrieved. The Number spacing [Y] command is used for setting the required number of spaces between characters in new or existing Text Items.

Module 3: Unit 12

Activity 12.3

Create a CAD drawing from the Co-ordinate File you created.
- Sheet centre: 8 455y 16 950x;
- Scale 1: 1 000; and
- A4 portrait.

Perform the following CAD functions explained above on this drawing:
- Line and Arc Drawing; and
- General Text Writing.

12.4 The SURPAC ENGINEERING Module Applications

A Demo version can be downloaded for SURPAC Software for Windows XP/Vista/7 from http://www.surpac.co.za/

SURPAC Surveying Software is structured to provide surveyors with a flexible selection of surveying software components.

The SURPAC Engineering Module includes the following applications (an excerpt from http://www.surpac.co.za/):
- Horizontal Alignment (Straights, Circular & Transition Curves);
- Vertical Curve Alignment and Profile Formations;
- Cross-Section Creating and Plotting;
- Longitudinal Section Creating and Plotting; and
- Sectional Volumes and Toe-Pegs.

12.4.1 Horizontal Curve and Straight Alignment

When given a set of user-defined parameters for a road or railway route that contains a series of horizontal Curves, this application will compute the [Y, X] (or [E, N]) Co-ordinates and Chainages of positions along the Centre Line of the route, or along a defined Offset Line.

The required Input parameters are:
- the [Y, X], or [E, N], Co-ordinates for the Start and End Points of the Section of the route to be computed, plus the Chainage of the Start Point. These Start and End Points represent the total Length of the Centre Line, or Offset Line, that are to be computed.
- the [Y, X] (or [E, N]) Co-ordinates for the Points of Intersection for all Curves that lie along the defined route (up to 100 Curves).
- the circular Radii for all Curves lying along the route;
- the Lengths of all Transition Curves, if any, for all the Curves along the route;
- the Chainage Interval required between Centre Line Points; and
- selection of Centre line or Offset Calculations.

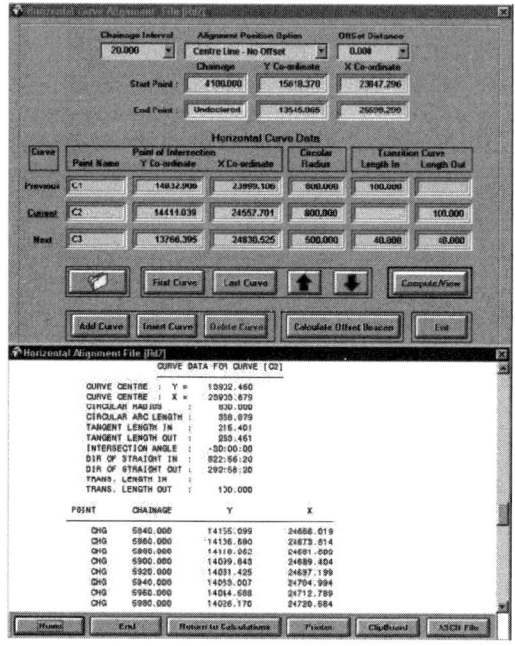

Fig 12.1 Horizontal curve and straight alignment

The route Section can also be as long, or as short, as required. The route Section may consist of a single Curve (where the Start and End values may be coincident with the BC and EC Points) or a single Straight. The maximum number of Points on any Centre Line File is unlimited.

If necessary, once a route or Section of a route has been computed, this application may be used for the editing of any of the Curve parameters. Curves may be Added, Inserted, Modified or Deleted from the original scheme. Any number of updates may be made and the Centre Line Co-ordinates recomputed and re-stored.

As the Centre Line values (Co-ordinates and Chainages) are computed they are stored on a user-defined Horizontal Alignment. Simultaneously, the Centre Line Point Values are displayed on the SURPAC printer emulation screen. From this screen, the output may be sent to the current Printer, to an ASCII File or to the Clipboard.

Setting out Data for Centre Line Points may be computed and printed via the Setting out Data Sheets.

The user may select to determine Centre Line values or Offset values on either side of the Centre Line. Offset lines are useful when curvilinear boundaries are required along an existing or proposed road/rail route.

Computed Points may be:
- converted to a User-defined SURPAC Co-ordinate File;
- plotted on a User-defined SURPAC CAD Plot Sheet;

- combined with SURPAC Vertical Curve Data for determining Section profiles; and
- used to generate Centre Line Setting out Field Sheets.

Vertical Curve Alignment and Section Formation Creation

This Program provides for the generation of the Design Cross-section Formation Data along a road or a railway route. This Data generation is completely dependent on the entered design criteria.

After entry of the required criteria, the Design Cross-section Formation Data are computed for each Section and stored in a User-defined Vertical Curve Alignment File. This information may be used by:
- the Cross-section Creation and Plotting;
- the Program for the plotting of Cross-sections;
- the Longitudinal Section Creation and Plotting for plotting Longitudinal Sections; and
- the Sectional Volumes and Toe-Peg Calculation Program for calculating Sectional volumes, Mass Haul volumes and Toe-Peg Distances for batter board placement.

The initial design criteria that are required by the Program are the design Data for the vertical Curves along the proposed route. For each Vertical Curve, the required Input Data are:
- the Intersection Point Chainage;
- the Intersection Point Elevation; and
- the Length of the Vertical Curve.

From this information, the Program will compute the parabolic vertical Curve parameters; the Start and End Chainages of the Curves; and the Gradients of the Straights connecting the Curves.

Any vertical Curve may be defined as having a Curve Length of 0 m. This is interpreted as being a 'kink', or bend Point along the Centre Line, and no parabolic Curve is computed. The two line gradients will intersect at the defined Chainage and at the defined Elevation.

The defined Point may merely be a chosen Point along a gradient; the incoming and outgoing gradients may be the same. This is useful for including 'odd' Chainage Points that may be required for Cross-section and/or Long Section plotting.

All vertical Curve Data should be entered to cover the total Length of the route under consideration. The design Centre Line elevations, at the designated Chainage Interval, are then computed.

After the Centre Line elevations have been computed, design Data for the Section Formations are entered, if they are required. If the Formation design criteria are constant along the entire route, then a single set of Input Data will be sufficient. If there are variations, such as the super-elevation that changes from Curve to Curve, then, separate design Data must be entered, Section by Section.

The Elements required for the Calculation of the design Formations are:
- the Start and End Chainages of the Section, or route;
- the median width and Slopes;
- the straight to Curve development Lengths and Ratios;
- the Curve widening within the Curve, if required;
- the Carriage Way cross-falls and super-elevations;
- the Carriage Way widths;
- the kerb Heights and Slopes;
- the pavement (shoulder) widths and Slopes; and
- any auxiliary Carriage Way widening along the straights.

The Program makes allowance for the above-mentioned design Elements for the Formation Cross-sections. These exclude the batter Slopes and side drains, which are defined at the Cross-section plotting stage. At the Cross-section plotting stage, designing is done with the Cross-section Creation and Plotting Program. At the volume Calculation stage, designing is done with the Sectional Volumes and Toe-Peg Calculation Program.

Once all Section design Data has been entered, the design Formations for each Section are also stored on the current Vertical Alignment File.

12.4.2 Cross-section Creation/Editing/Plotting

Cross-sectional Data may be generated through a variety of different methods, namely:
1. By manual (Keyboard) entry.
2. By Interpolation when combining a Horizontal Alignment (Centre Line) File and a Tacheometric File.
3. The use of a Horizontal Alignment File infers that the Centre Line Points are at regular Intervals, and will lie along mathematically correct geometric entities such as straight lines, circular Curves and Transition (clothoid) Curves.
    - The two defined files must have a common area of overlap. Further information such as the required Section Chainage range, the Section widths and interpolation Distances along the Sections may be entered.
    - This method becomes a technique for creating Cross-sections without having to stake the Centre Line, or physically measure the Sectional Data. This technique is not suitable, however, for terrain that contains man-made features, such as drains, existing roads and the like.

Module 3: Unit 12

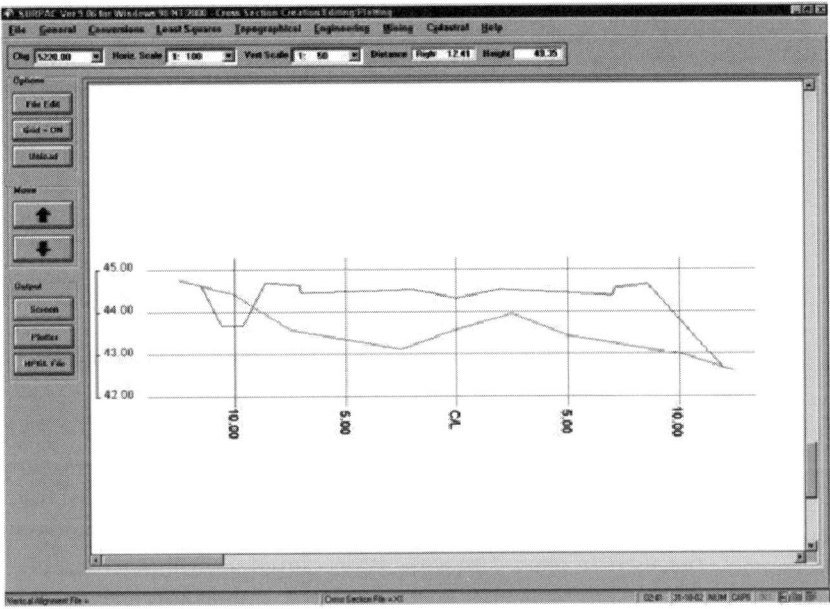

Fig 12.2 Cross-sections

4. By Interpolation, when combining a Co-ordinate File and a Tacheometric File:
   - The use of a Co-ordinate File accommodates the Centre Line Points that need not occur at regular Interval and may not necessarily lie along mathematically correct geometric entities.
   - The two defined files must have a common area of overlap. Further information such as the required Section Chainage range, the Section widths and interpolation Distances along the Sections may be entered. Interpolation is carried out via a least squares plane fitting technique.
   - As for the previous method, this method becomes a technique for creating Cross-sections without having to stake the Centre Line or physically measure the Sectional Data. This method is not suitable, however, for terrain that contains man-made features such as drains, existing roads and the like.
5. By using a Horizontal Alignment (Centre Line) File plus extraction of BreakLine Data from a Tacheometric File:
   - This option searches through a User-defined Tacheometric File for all BreakLine information and then combines this information with Centre Line information that is taken from a User-defined Horizontal Alignment File.
   - The two defined files must have a common area of overlap.
   - Only those Points that represent intersections between a BreakLine and defined Cross-sections will be Added to the

233

Cross-section File. Therefore, by manipulating BreakLines in a Tacheometric File it is possible to pre-select features that will be reflected on the Cross-sections. No interpolation will take place using this Option.
- This method becomes a technique for creating Cross-section information on discreet topographical features, without having to stake the Centre Line or physically measure the Sectional Data. This technique is suitable for both natural terrain and terrain that contains man-made features such as drains, existing roads and the like.

6. Generate Cross-sections using both Interpolation and BreakLine Data from a Tacheometric File:
   - This option uses the Centre Line Data that is taken from a User-defined Horizontal Alignment File to position and orientate the Cross-sections. It then searches the defined Tacheometric File for all BreakLine Information and combines this information with least squares Interpolated Heights to supply the most complete Cross-section information of all the options.
   - The resulting Sections are more comprehensive than those of the previous options are, and therefore take somewhat longer to generate.
   - This method becomes a technique for creating Cross-section information on discreet topographical features, plus Interpolated Heights, without having to stake the Centre Line or physically measure the Sectional Data, and is suitable for both natural terrain and terrain that contains man-made features, such as drains, existing roads and the like.

7. Load Cross-section Information held in a Tacheometric File:
   - This option combines information that is held in a User-defined Horizontal Alignment File and surveyed Sectional information held in a User-defined Tacheometric File. The two files must have a common area of overlap.
   - Further information such as required Section widths and acceptable Off-line tolerances along the Sections must be entered.
   - This method becomes a technique for generating Cross-sections that have been Surveyed using polar techniques and stored in, and then transferred from, an electronic logger or Total Station. This method requires that the Centre Line is staked and the Sections are measured in the field and is, therefore, suitable for terrain that contains man-made features.

8. Load Cross-section Data from a Handi-Data 'LEVELS' format ASCII File:
   - When using this option, the Cross-section Data are read from an ASCII File, set in the 'Handi-Data Systems SURPAC' format. This method provides direct reading and loading of Data as measured

in the field. For further information on the ASCII File format, refer to the 'LEVELS' User's Guide for the PSION Organiser, or for the PSION Workabout 'DISK Transfer'.

9. The total number of Points per Cross-section is 51: the Centre Line Point plus 25 Points left of the Centre Line and 25 Points right of the Centre Line.
10. Cross-sections may be Displayed on the Screen, and are directly Plotted to a Printer/Plotter or stored in an HPGL File format.
11. Cross-sections may be combined with a Vertical Curve Alignment File Data that are created by the Vertical Curve Alignment Program. If Design File information is included, the Program allows for the inclusion of User information that is related to the construction of batter Slopes, drainage channels and the like. This combination is used both for the plotting of Cross-sections and for the generation of Sectional Volumes.

12.4.3 Longitudinal Section Creation/Editing/Plotting

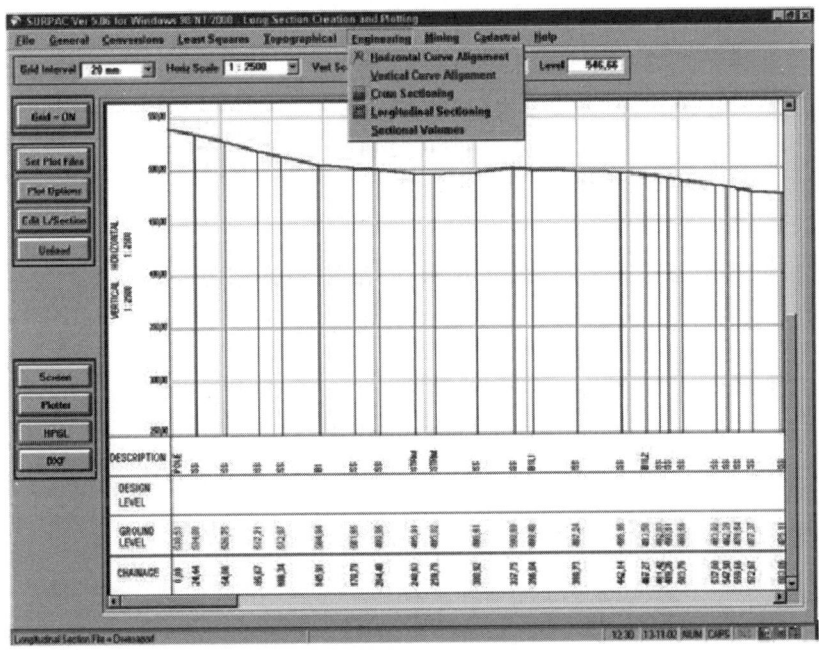

Fig 12.3 Longitudinal sections

The Data used for display, or plotting, may be any of the following:
1. Section Plotting Using both Design and Cross-section Files:
   - This combines the Centre Line of a Vertical Curve Alignment File and User-designated Elements of a Cross-section File. The selected

Element may be any of the available 51 Elements of a Cross-section. These 51 Elements consist of 25 Points left of the Centre Line, the Centre Line and 25 Points right of the Centre Line.
2. Section Plotting Using a Cross-section File only:
   - This uses any User-designated Element of the available 51 Elements of a Cross-Section File.
3. Section Plotting Using both Design and Long-Section Files:
   - This combines the Centre Line of a Vertical Curve Alignment File and a Longitudinal Section File (created through this application — see further on).
4. Section Plotting Using a Long-Section File only:
   - This uses a Longitudinal Section File, as created and/or changed through this application.
5. Section Plotting Using a Design File only:
   - This uses a Centre Line of a Vertical Curve Alignment File.
6. Section Plotting of Random Grouped Section Files:
   - This option displays, or plots, a series of Long-Section Files (known as Random Grouped Files), which have been generated from Data in a Tacheometric File, and in which the Points follow a specific numbering coding system used by SURPAC.

The Options for Creating a Longitudinal Section File:
1. Generate Sections from a Tacheometric File and defined Bend Points:
   - This method locates and sorts Long-section Data along a defined Section Line, or a series of consecutive Section lines within a Tacheometric File. The Section Line positions are defined by identifying consecutive Bend Points that are held in the Co-ordinate File. The two defined files must have a common area of overlap. This is not an interpolation technique; and only surveyed Data will be used. This method is best suited for measurements downloaded from a total station or logger. Side Slope Points may also be included.
2. Interpolate Sections from a Tacheometric File and defined Bend Points:
   - This method uses a system of Least Squares plane fitting, along the User-defined Section Line(s) and at a defined Interval. The Section Line positions are defined by consecutive Bend Points, which are identified and held in the Co-ordinate File. The surveyed Data are Tacheometric Data (plus BreakLines, where necessary) that are held in a User-defined Tacheometric File. This method is best suited for determining random Sections that occur over a surveyed area where no direct long-section information has been surveyed. This method becomes a technique for creating

Long-Section information on discreet topographical features (and Interpolated Heights) without having to measure directly along the Section Line. This technique is suitable for both natural terrain and terrain that contains man-made features, such as drains, existing roads and the like.

3. Create Sections from a Contour Triangulation File plus defined Bend Points:
   - This method searches for all possible intersections between the Longitudinal Section Lines (defined by the Bend Points) and the sides of Triangles (as created in a User-defined Contour Triangulation File). The Section Line positions are defined by consecutive Bend Points that are held in the Co-ordinate File. This method is best suited for determining random Sections that occur over a surveyed area where no direct Long-Section information has been surveyed. It provides Long-Section Data that are based on Contour Triangulation, without having to measure directly along the Section Lines. This technique is suitable for both natural terrain and terrain that contains man-made features such as drains, existing roads and the like.

4. Load Section Information from a Handi-Data 'LEVELS Lng-Sect' File:
   - This method will load Long-Section Data from a Handi-Data Systems PSION Organiser, or PSION Workabout, 'LEVELS' Lng-Sect File. It will do this as long as the Data are captured in SURPAC Format. For further information, refer to the 'LEVELS' User's Guide for the PSION Organiser, or the PSION Workabout, DISK Transfer.

5. Load Long Section Information from an ASCII File:
   - Data are imported from a Fixed Column ASCII File.

6. Create a series of Grouped Random Section Files from a Tacheometric File:
   - When using this function, it is possible to rapidly create a series of Long-Section Files from Data that are held in a Tacheometric File. The Points for each Section that is created are identified by means of the Point numbers of the Data in the Tacheometric File, using a number coding system. Each Section created will be in its own separate Longitudinal Section File. Each Longitudinal Section File will be given the Name of the Tacheometric File plus the Section Number.

12.4.4 Sectional Volumes

This application combines a Vertical Curve Alignment File and a Cross-section File to determine Sectional volume information.

The following pre-conditions must exist:
- The two files selected must have an overlapping strip that contains Sections with common Chainages.
- The Vertical Curve Alignment File must contain the Design Formation Cross-sections, and not just the Vertical Curve information.
- The Cross-section File must contain ground, or surveyed Sections wide enough to cover the Design Formation Cross-sections, plus the Batter Slopes and side drains.
- If the extent of a ground Section, or a surveyed Section, is not wide enough to accommodate the design criteria mentioned above, the Program will connect the last Design Formation Cross-section Element (such as the End of pavement) with the last surveyed Section Point. This function can be put to good use when a section of road that has a retaining wall, for example, does not follow the normally defined Batter requirements. In this case, you should limit the relevant survey Cross-sections to the wall position.

Before computing any volumes, the application will prompt you for information that is related to the method of connection between the Design Cross-sections and the Surveyed Cross-section. This information consists of:
- the Start and End Chainages of the required strip;
- the angles for Cut Batters and Fill Batters;
- the Drainage Channel information — this is only applicable to Sections in Cut;
- the Depth of Top Soil Stripping. This value, if required, will be applied to all surveyed Section Heights;
- the Depth for Special Grade Material. If a value is given, the amount given will be subtracted from the Design Section Carriage Way Heights. This is done to compensate for the 'boxed out' volume that is reserved for the Special Grade Material; and
- the Bulking/Formation Factor. This factor is used to compensate for the difference in compaction between natural material, Cut material and Filled material. If you define a factor greater than 100%, this implies that the Cut material will occupy a larger volume when used for Fill.

Finally, you have the choice of computing and displaying the Accumulated Cut and Fill Volumes for the strip defined, or of computing and displaying the Cross-sectional Cut and Fill Areas for each Cross-section.

Once all the required parameters have been set, click the Compute Button. The Program will compute the volumes and display the output

in the SURPAC Printer emulation Screen. The application output Data consist of :
- the Section Chainages;
- the Cut and Fill Volumes between two consecutive Cross-section;
- the Mass Haul, or Net Cut and Fill Volumes, up to the current Cross-section;
- the Accumulated Cut and Fill Volumes up to the current Cross-section;
- the Cross-section Cut and Fill Areas for the current Cross-section; or
- the left and right Toe-Peg Distances for the current Cross-section.

Appendix

Semi-circular hooks

Shape codes 32, 33 and 72

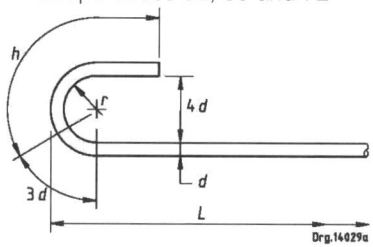

Cutting length of bar equals L + h
Bends that form end anchorages
Shape codes 34, 35 and 42

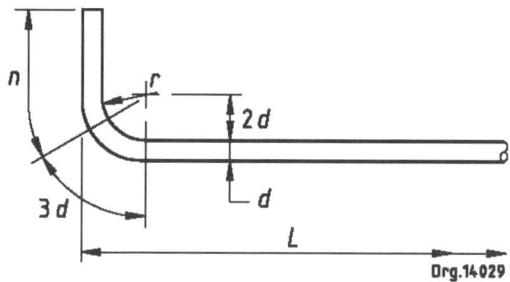

Cutting length of bar equals L + n

Dimensions in millimetres

1	2	3	4	5	6	7	8	9	10
Nominal size of bar d	6[1]	8	10	12	16	20	25	32	40
Hook allowance h	100	100	120	120	160	200	260	320	400
Bend allowance n	100	100	100	100	100	120	160	200	240
Radius r	12	16	20	24	32	40	50	64	80
1) Non-preferred.									

Figure A.1 Minimum hook, bend and radius allowances for hot-rolled mild steel bars that comply with SANS 920

Appendix

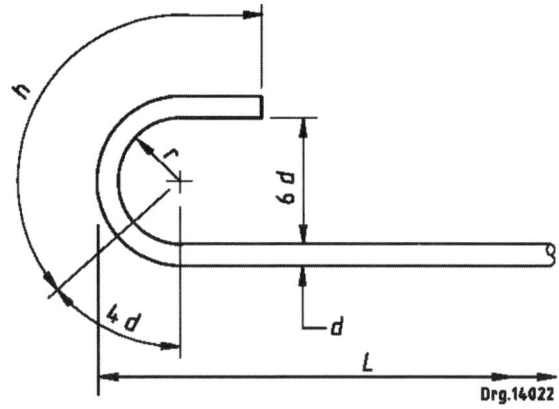

Cutting length of bar equals L + h

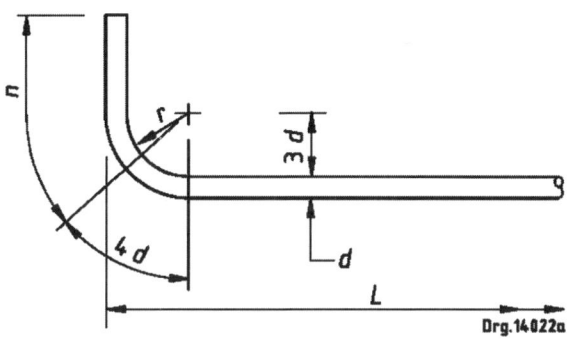

Cutting length of bar equals L + n

Dimensions in millimetres

	1	2	3	4	5	6	7	8	9	10
Nominal size of bar d		6[1])	8')	10	12	16	20	25	32	40
Hook allowance h		100	100	120	160	200	240	300	400	480
Bend allowance n		100	100	100	100	120	140	180	220	260
Radius r		18	24	30	36	48	60	75	96	120
1) Non-preferred.										

Figure A.2 Minimum hook, bend and radius allowances for high yield stress and cold-worked steel bars that comply with SANS 920

Table A.1 Measurement of bending dimensions

1	2	3	4
Shape code	Method of measurement of bending dimensions	Calculated length[1] (measured along centre line)	Information (sketch and dimensions) to be given in bending schedule
20	*A* — Drg.14304a	A	Straight
32	*A* — Drg.14304b	$A + h$	*A* — Drg.14304h
33	*A* — Drg.14304c	$A + 2h$	*A* — Drg.14304i
34	*A*, *n* — Drg.14304d	$A + n$	*A* — Drg.14304j
35	*A*, *n*, *u* — Drg.14304e	$A + 2n$	*A* — Drg.14304k
36	*A, B, C, D, E* — Drg.14304f	$(A + C + E) + 0{,}57(B + D) - 3{,}14d$ If B or $D > 400 + 2d$, see NOTE 1 of 4.2.5	*A, B, C, D* — Drg.14304l
37	*A, B, r* — Drg.14304g	$A + B - \tfrac{1}{2}r - d$ If r is non-standard, use shape code 51	*A, B* — Drg.14304m
38	*A, B, C, r* — Drg.14024a	$A + B + C - r - 2d$ $A + B + C - r - 2d$	*A, B* — Drg.14024b
39	*A, B, C* — Drg.14024c	$A + 0{,}57 B + C - 1{,}57 d$ If $B \geq 400 + 2d$, see NOTE 1 of 4.2.5.	*A, B* — Drg.14024d

Appendix

Table A.1 (continued)

1	2	3	4
Shape code	Method of measurement of bending dimensions	Calculated length[1] (measured along centre line)	Information (sketch and dimensions) to be given in bending schedule
41	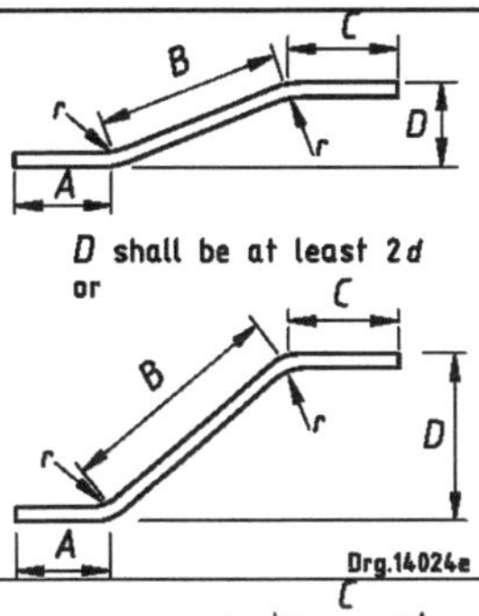 Drg.14024e — D shall be at least $2d$	If angle with horizontal is 45° or less $A + B + C$ otherwise see NOTE 4	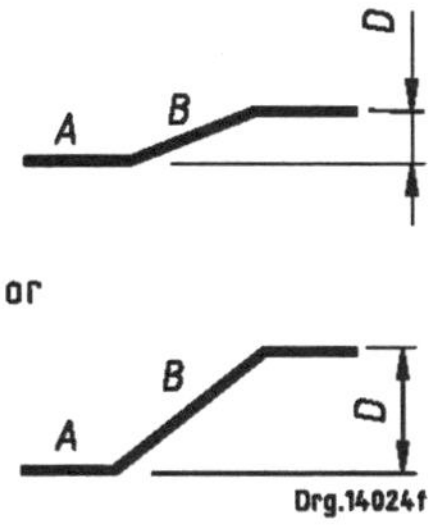 Drg.14024f
42	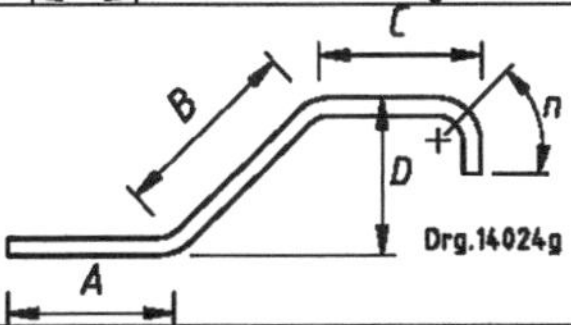 Drg.14024g	If angle with horizontal is 45° or less $A + B + C + n$ otherwise see NOTE 4	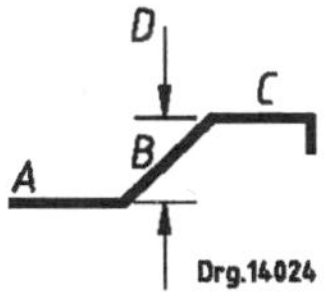 Drg.14024
43	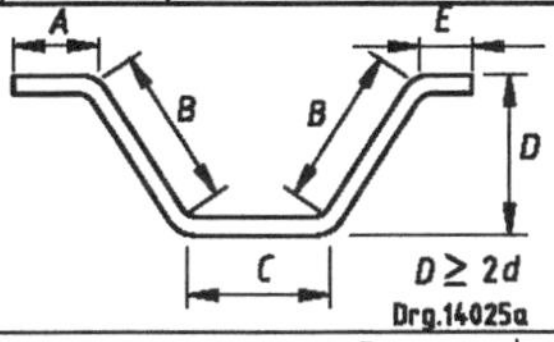 Drg.14025a — $D \geq 2d$	If angle with horizontal is 45° or less $A + 2B + C + E$ otherwise see NOTE 4	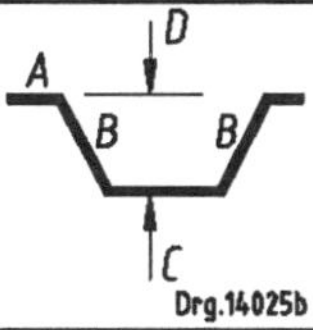 Drg.14025b
45	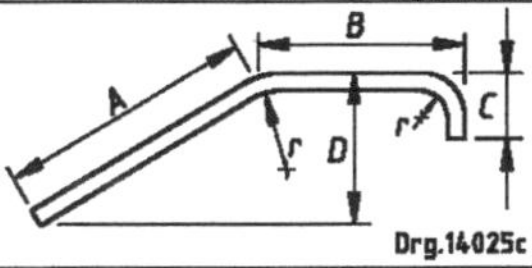 Drg.14025c	If angle with horizontal is 45° or less $A + B + C - \frac{1}{2}r - d$ otherwise see NOTE 4	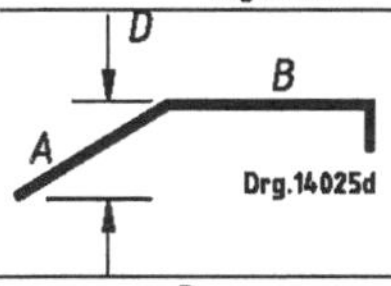 Drg.14025d
48	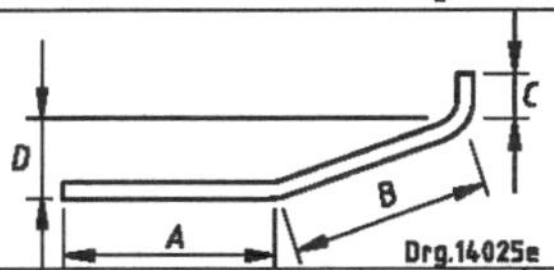 Drg.14025e	If angle with horizontal is 45° or less $A + B + C$ otherwise see NOTE 4	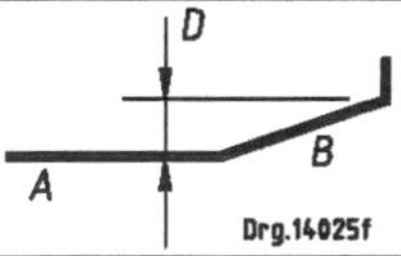 Drg.14025f
49	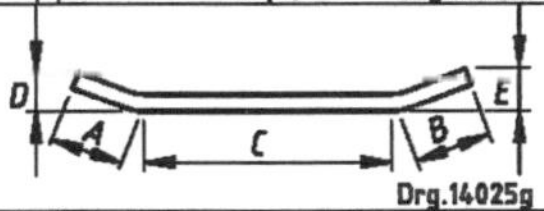 Drg.14025g	If angle with horizontal is 45° or less $A + B + C$ otherwise see NOTE 4	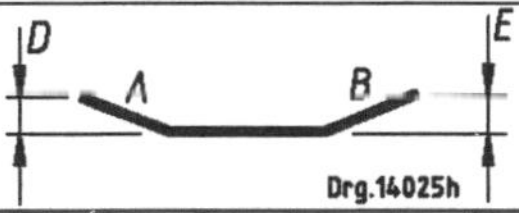 Drg.14025h
51	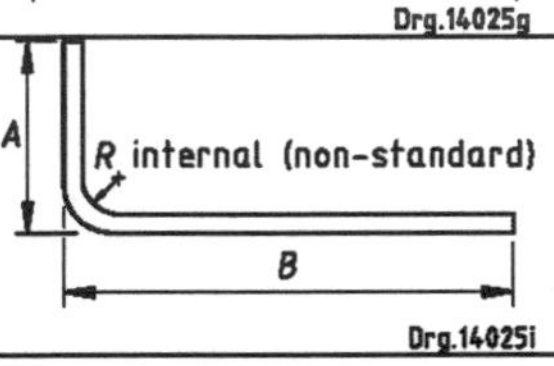 R internal (non-standard) Drg.14025i	$A + B - 0{,}43\,R - 1{,}21\,d$ If R is standard, use shape code 37 If $R \geq 200$ mm see NOTE 1 to 4.2.5	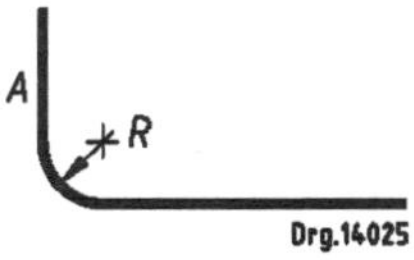 Drg.14025

Amdt 1

Table A.1 (continued)

1	2	3	4
Shape code	Method of measurement of bending dimensions	Calculated length[1] (measured along centre line)	Information (sketch and dimensions) to be given in bending schedule
52	(Drg.14026a)	$A + B + C + D - {}^{3}/_{2}r - 3d$	(Drg.14026b)
53	(Drg.14026c)	$A + B + C + D + E$ $-2r - 4d$	(Drg.14026d)
54	(Drg.14026e)	$A + B + C - r - 2d$	(Drg.14026)
55	(Drg.14027a)	$A + B + C + D + E$ $-2r - 4d$	(Drg.14027b)

Table A.1 (continued)

1	2	3	4
Shape code	**Method of measurement of bending dimensions**	**Calculated length[1] (measured along centre line)**	**Information (sketch and dimensions) to be given in bending schedule**
60	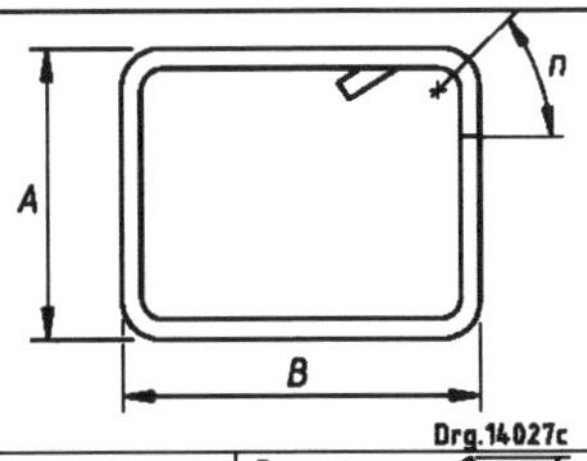n A B Drg.14027c	2*A* + 2*B* + 2*n* - 3*r*/2 - 3*d*	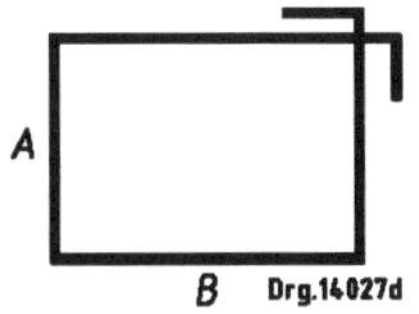A B Drg.14027d
62	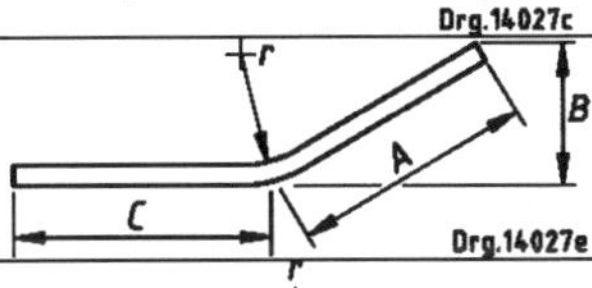r B A C Drg.14027e	If angle with horizontal is 45° or less *A* + *C* otherwise see NOTE 4	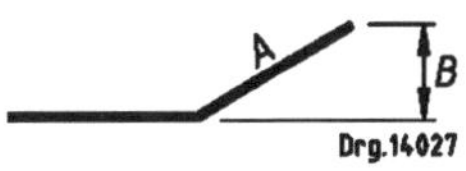A B Drg.14027
65	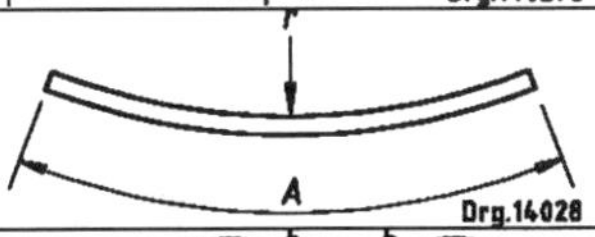r A Drg.14028	*A*	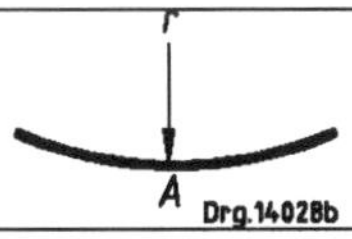r A Drg.14028b
72	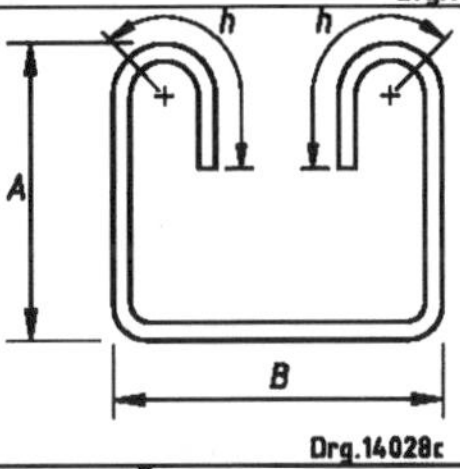h h A B Drg.14028c	2*A* + *B* + 2*h* - *r* - 2*d* See NOTE 3	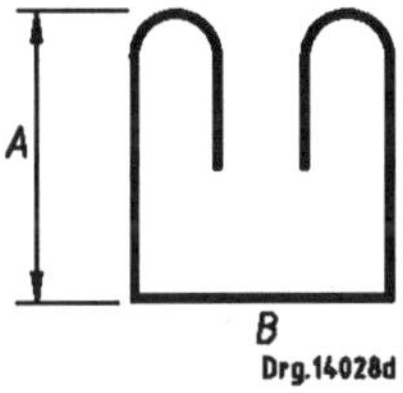A B Drg.14028d
73	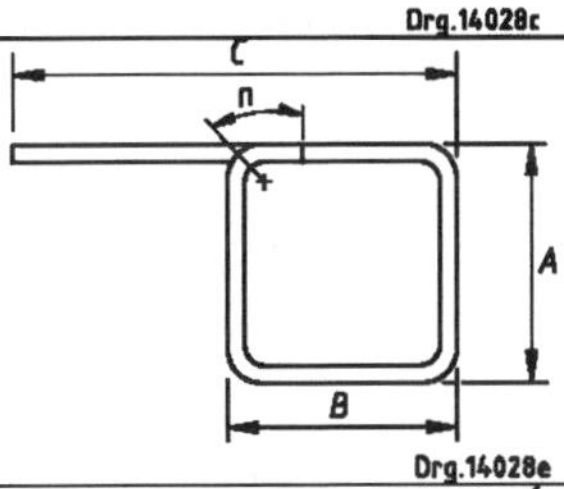C n A B Drg.14028e	2*A* + *B* + *C* + *n* - 3*r*/2 - 3*d*	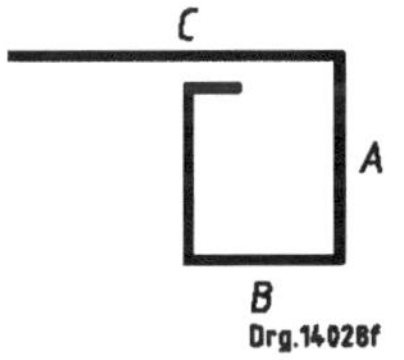C A B Drg.14028f
74	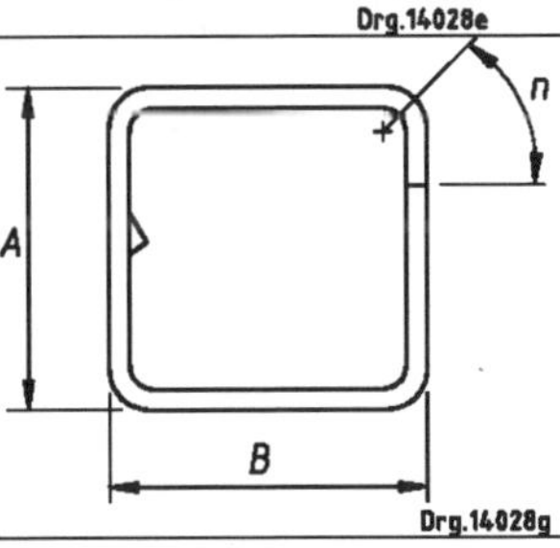n A B Drg.14028g	2*A* + 3*B* + 2*n* - 2*r* - 4*d*	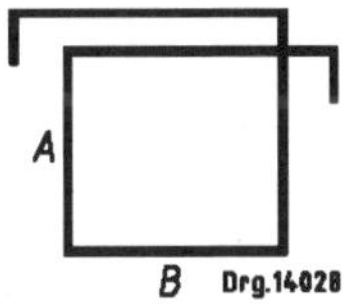A B Drg.14028

Drawing for Civil Engineering

Table A.1 (continued)

1	2	3	4
Shape code	Method of measurement of bending dimensions	Calculated length[1] (measured along centre line)	Information (sketch and dimensions) to be given in bending schedule
75	Drg.14030a	$A + B + C + 2D + E + n\text{-}5\,r/2 - 5d$	Drg.14030b
81	Drg.14030c	$2A + r + d + 2h$ For larger radii, see 4.2.5 Note to table 5 and refer to shape code 99	Drg.14030d
83	Drg.14030e	$A + 2B + C + D - 2r - 4d$	Drg.14030f B is overall dimension
85	$C \geq 2r + 2r$ Drg.14030g	$A + B + 0{,}57C + D - \tfrac{1}{2}r - 2{,}57\,d$ If C is greater than $400 + 2d$, see 4.2.5	Drg.14030
86	Drg.14021	$\dfrac{C}{B} \times \pi \times \sqrt{(A-d)^2 + B^2} + X$ Where $X = 2 \cdot \pi \cdot (A-d)$ and B does not exceed A/5 or 150 mm, whichever is the least. There shall be at least two full turns in the helix.	Drg.14021b Where an additional turn is required at the helix' end(s), it shall be treated as a shape code 99.

Amdt 1

Appendix

Table A.1 (continued)

1	2	3	4
Shape code	Method of measurement of bending dimensions	Calculated length[1] (measured along centre line)	Information (sketch and dimensions) to be given in bending schedule
99	All other shape codes		Provide either a dimensioned sketch or a referenced sketch with dimensions A, B, C, D, EIR shown in the schedule. The dimension that is to allow for the permissible deviation shall be indicated, as shown in the relevant cases in column 4 of this table.

Where so desired, the standard radius r for corners may be increased, but shall not apply to end anchorages denoted by h or n. See 4.2.3.1.

NOTE 1   r indicates the minimum values in tables 3 and 4.
NOTE 2   The dimensionless sections of the bars are the run-off legs.
NOTE 3   Ensure that dimension B is greater than $14d$ for smooth mild steel and greater than $18d$ for deformed steel.
NOTE 4   The length formula is approximate and when bending angles exceed 45°, the length should be calculated more accurately allowing for the difference between the specified overall dimensions and the true length measured along the centre line of the bar or wire.

1) These lengths are approximate and generally involve a slight positive difference from the exact mathematical expressions.

Acknowledgements

The author and publisher would like to thank the following for their permission to reproduce material in this book:
- The Architectural Press
- Blackwell Science Ltd
- The Institute of Topographical and Engineering Surveyors of South Africa
- Mr L.V. Leech
- Macmillan Press Ltd for permission to reproduce diagrams from *Structural Detailing for Architecture, Building, and Civil Engineering* by Peter H Newton
- Murray & Roberts Gills Mason
- John Newman of Teaching and Learning Enterprises Ltd (TALENT) for use of COMPACT program
- The South African Bureau of Standards and George Pauer for use of SANS 282 Edition 5.1 of 2004
  Disclaimer: The South African Bureau of Standards accepts no liability for any damage whatsoever that may result from the use of the said standards or the information contained therein, irrespective of the cause and quantum thereof. Extracts from SANS 282:2004 appear in accordance with permission granted by the South African Bureau of Standards.
- The Southern African Institute of Steel Construction and Spencer Erling, SAISC Director, for the use of *The Blue Book: Structural Steel Tables*. Available at www.saisc.co.za

The publishers have made every attempt to trace and contact copyright holders. If any copyright infringement should have occurred, please inform the publishers so that the error can be rectified in the next edition.

Scale drawings

Throughout the text the author refers to a number of scale drawings, which appear in the pocket at the back of the book. To help you find the drawings easily, please refer to the next two pages, which show you how they have been placed on the sheet at the back of the book.

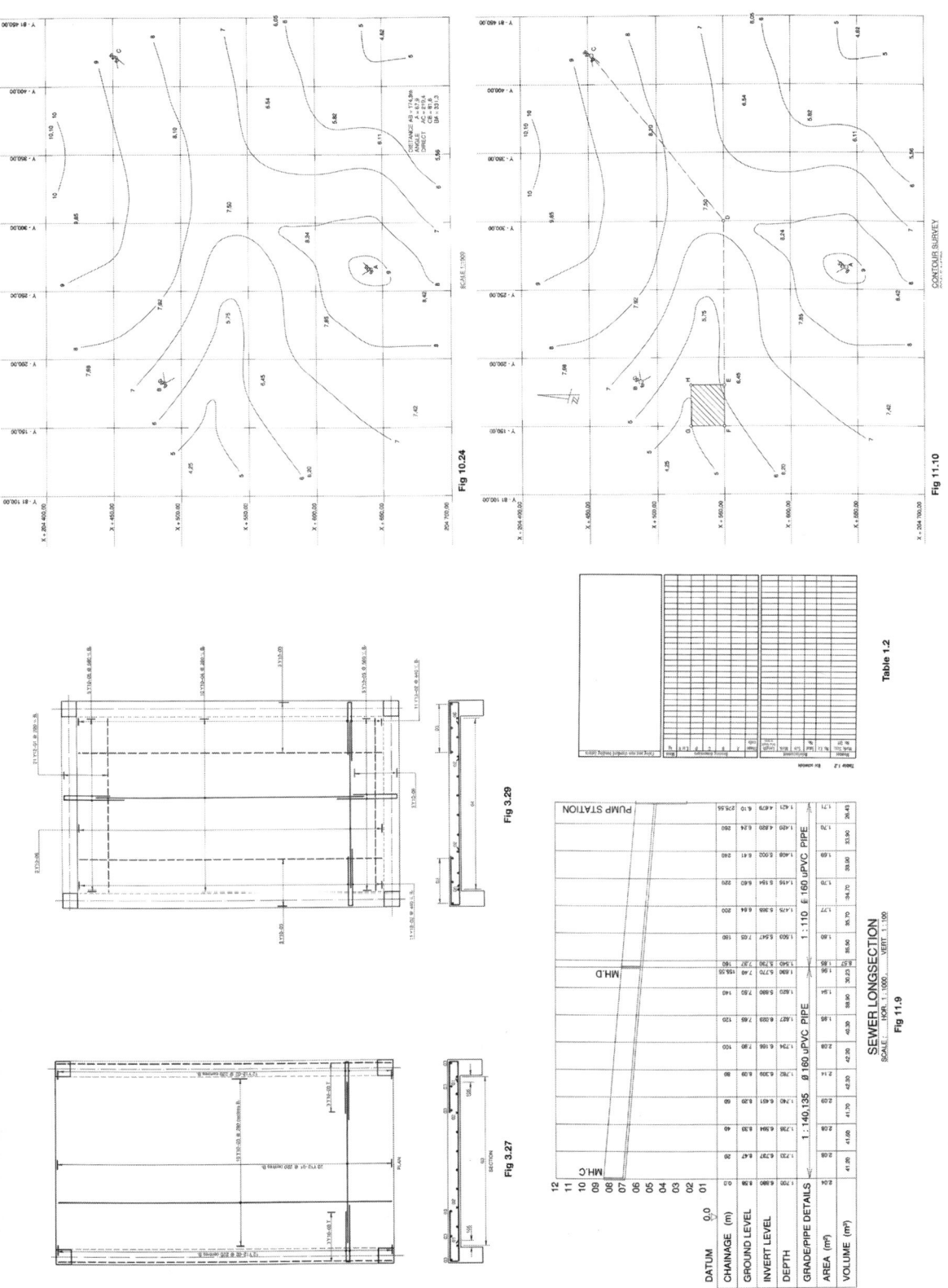

249

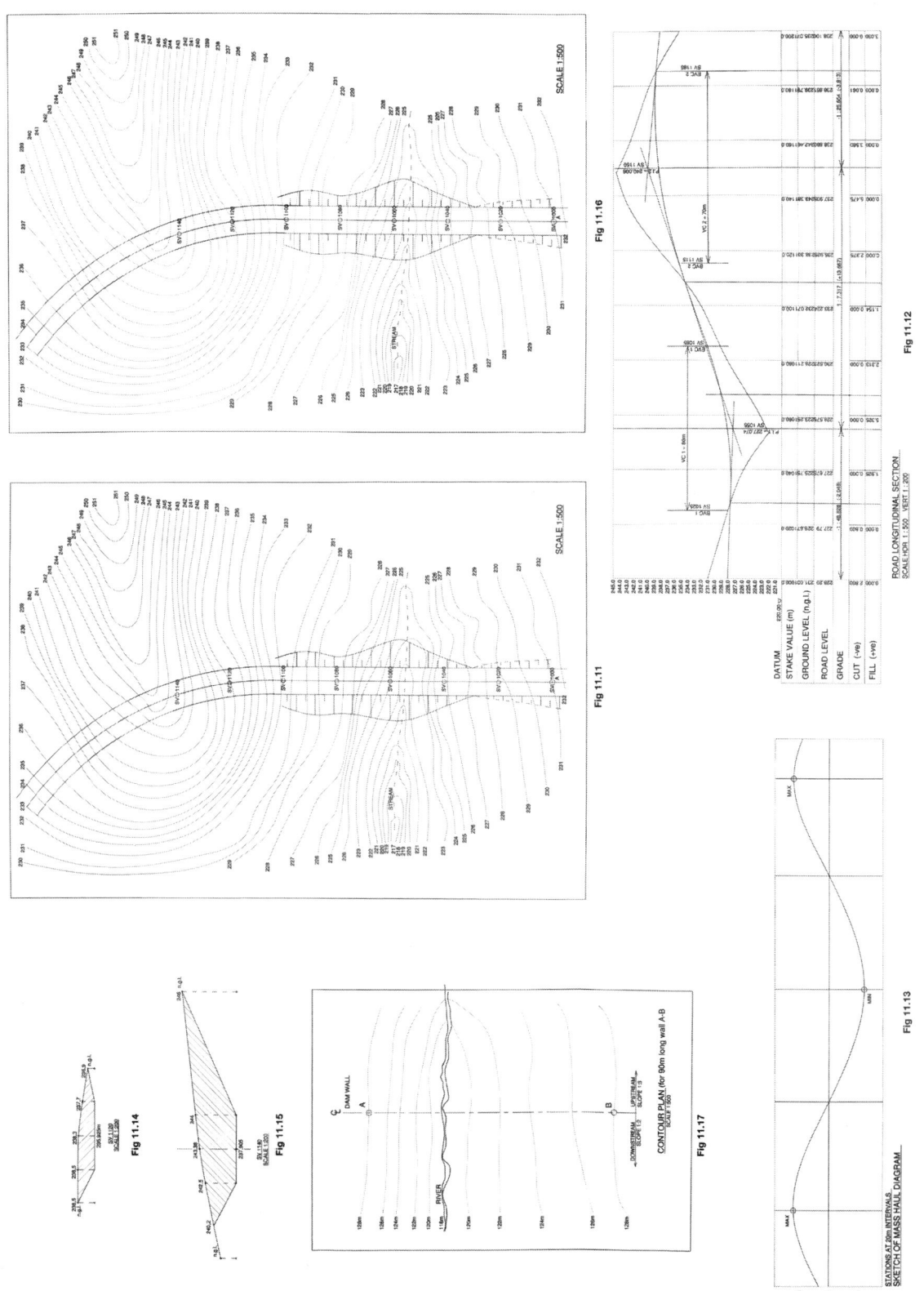

Index

A
aggregate 2, 4
angles, equal leg *112-115*
　　unequal leg *116-117*
arc drawing 226
auto-compilation 221-223
average haul distance 180
axes 181
axial uplifting forces 98, 122

B
backmark 118-120
balance line 206
bar marks 14, 40, 42, 57, 59-62, 66
bar schedules 12-15, *13*
bars, calling up of 26
bases, concrete *see* concrete bases
bases, steel *see* steel bases
beams, concrete *see* concrete beams
beams, steel *see* steel beams
bend points 236, 237
bending *8, 9*, 98
bending dimensions 14, *242-247*
bending moment 38, *58, 60*, 122
bending schedules 12-15, *30, 33*, 60, *63, 74, 75-77*
bent bars *39*
bent-up bars *44*
blinding concrete 9
bolt pitches 135
bolted connections 101-118
bolted truss detailing *157, 159*
bolts 101
　　layout of *118*
　　vs welding 155
bond *3*
borrow pits 203
breakline data 233-234, 236
bulking 205
bulking formation factors 238

C
CAD 225-227
　　erasing 227
　　text writing 228

cadastral data 180
　　plotting *199-200*
cadastral symbols 186-*188*
cantilever footings 17
carriage way heights 238
cement 2, 4
　　vs steel 4
centre line elevations 231
chairs 47-*48*,
channels *110-111*
cleats 98, *136*, 149
columns, concrete *see* concrete columns
columns, steel *see* steel columns
combined footings 17, 21-24, *22, 31*
　　example 29-33
COMPACT 83-95
compression 4, 98
　　direct 9
computer aided concrete training *see* COMPACT
computer-aided drawing *see* CAD
concrete 2
　　outline 26
　　reinforced see reinforced concrete
　　tension vs compression 8
concrete bases 17
　　reinforcement of 9
concrete beam and slab construction 36, 49, 50, *51*, 70-71, *72, 75*
　　steel detailing *81*
concrete beam-column *see* concrete column-beam
concrete beams 5, 36-38
　　detailing of 38-44, *41*
　　supports of *42*
concrete column-beam junctions 50, *52, 53*, 54
concrete columns 5, 9, 20-24
　　detailing of *19*-20
　　failure modes of *9*
　　junctions between *11*
　　load 21
　　reinforcement of *10*, 11, *56, 57*
concrete cover 2
concrete slabs 5, *44-46, 42*

251

construction joints 11, *12*
continuous beams 36-38, *38*, 39-40
    reinforcing for *40*
contour plan 199-200, 212
contour triangulation files 237
contour values *185*
contours 180, 183-188, *185*, *186*
contra flexure 38
co-ordinate files, 220-221, *222*, 233
    printing 223-224
co-ordinates 181, 189
cross steel 22
cross-section files 235-236
cross-sections 40, 180, 204-205, *216*
    creation of 232-235, *233*
    design formation 238
    loading data 234, 237
curve alignments 229-232, *230*
cut and fill 203-204, 238-239
cutting lists 12

D
dampness, protection against 9
design speed 204
dimension lines 19, 26
dispersion angles 2, 22
distribution steel 47
drainage channel information 238
drawing, general principles of 8-12

E
earthworks 203
eccentric connections *141*, 142
edge distance *101*
electrodes 98, 102
elevation 40
end area method 180, 205
end command 226
end distance *101*
end plates 149
engineering format beacon description list 221-223
equal leg angles *see* angles, equal leg
equator 180, 182
excavation 203

F
floor and beam detail *41*
floor plans, steel *152*
floor slabs 55
    detailing of 47-49, 64-67, *65*
    reinforcing for *49*
footings 17, 20-24

formation grade line 206
formwork 11
foundations 9, 17-18
freehaul 180, 206, 208, 209

G
gauge distance 118, 135
grade points 206
grid lines 180, 181-*182*, 189
    methods of plotting *189-197*
grid system 131, *134*
grid values 189
ground pressure *21*, 22
grout 98, 130
gussets 98, 129, *130*

H
hand-flame edges 98, 101
Handi-Data 234-235, 237
haul 180, 203
haul distance 180, 206, 208, 209
heel and toe of I-section 119
high yield steel 2
holding down bolts 122-*123*
holes/holing 98
    positioning of *118*, 120
    diameter of *122*
    dimensioning of 120, *121*
horizontal alignment files 232-233
horizontal curve alignments 229-231, *230*
H-sections, parallel flange *108-109*

I
ingots 98
interpolation 180, *183*-184, 198, *199-200*, 233, 236
I-sections, parallel flange *102-105*
I-sections, structural detailing of *119*
I-sections, taper flange *106-107*
isolated footings 17, *20-21*, 26
    example 24-29

J
join calculations 224

K
kickers 19

L
lacing bars 74-75
lap lengths 2, *3*
lapping 11
lattice girders *160*-164, *170*, 172

bolted *161, 164*
welded *162, 163, 170, 172*
layers 26-27, 49
least squares 236
limit of economical haul 180
line drawing 226
links 5, 10, 19, *20*, 38-*39*
long span trusses *155*
longitudinal bars 19
longitudinal reinforcement 10
longitudinal section lines 237
longitudinal sections 180, *210*, 212
    creation 23, *235*-239
longitudinal steel 22
long-section files 236, 237

M
mansard-type trusses *155*
mass 14
mass haul diagram 203-206, *205, 207*, 211
    balancing procedure 206-207
members, definition of 12
meridians 180, 189
minimum pitch *101*
moment 98

N
natural ground 206
node points *161, 162*
north point 189

O
one-way supported slabs 45
orientation 189
origin 180, 181
overhaul 180, 206, 208

P
panel points *154*
parallel flange channels *110,111*
parallel flange H-sections *108-109*
parallel flange I-sections *102-105*
pile footings 17
pitch 98, 101
plotting methods 189-197
plotting sequences 197-199
plotting sheets 225-227
polar calculations 223-224
portal frames *165, 166, 167, 168, 171*, 172, *176, 177*
purlins *154, 169*

R
raft footings 17
random grouped section files 236
reinforced concrete 1, 4-5
    drawings for 15
reinforcement 2, 4, *23*, 49
    detailing of 5-8
    spacing 7
    standard dimensions of 15
rivets 98
    vs welds 155
roof systems 169-171
roof trusses *153*-159, *154, 155, 156, 157*
    steel *173-174*
    vs lattice girders 160
roofs 153

S
SA Bureau of Standards 8
scale 189
    manipulating 40, 42
seating cleats 149
section line positions 237
sectional volumes 231, 237-239
semi-circular hooks *240-241*
sequential system 146
shape codes 14
sheared edges 98, 101
shearing 4, *8, 9, 44*
shrinkage 205
shrinkage-temperature bars 49
shuttering *see* formwork
simply supported beams *37*
simply supported slabs *65, 70, 75, 79*
site plans 203-207
slab and beam *see* beam and slab
slabs, concrete *see* concrete slabs
snap radius 226-227
spacer bars *42*
spacers 9
span 98
splices/splicing 131-135, *136*
    beams 146, *149*
    plates 98, 135
spoil heaps 203
spot-heights 180, 183, *184*
    plotting 183, 198, *199-200*
spread footings 17, *18*
stanchions 98, 131
start command 226, 227
starter bars 21
steel
    high yield *see* high yield steel

mills 98
types of 100
vs concrete 4
steel bars, fully stressed 3
steel base-foundation contact 129-*130*
steel bases, as part of structural steelwork *129*-130
steel beams 138, 146
    connections to beams 146-*151*, *150*
    connections to columns *139*, *140*, 142, *143*, *145*
    connection to corner stanchions *140*
    details *147*
    dimensioning *148*
    eccentric connections *141*, 142
    notching of 138-142, *141*
steel columns 130-*132*, *137*
    connections to beams *139*, *140*, 142, *143*, *145*
    details *133*
steel frame buildings 99
stirrups 2, 4, 5, 28-29, 38-39, *42*, *43*, 60
straight curve alignments 229-231, *230*
structural steel 100
structures, positioning of on plans 203-207
supported slabs 45-*46*, 65
SURPAC 220-225
    line styles 227
    module applications 229-230
surveying 180, 181
symbols
    cadastral *186-188*
    for welds *124*, 125-126
    location of bars 7
    rivets and bolts 121

T
tacheometric files 232-235, 236, 237
taper flange channels *110*

taper flange I-sections *107*
tensile failure 4
tensile strength of steel 4
tensile zone 39-*40*
tension 4, 21, 22, 98
tension lap 59
toe-peg calculations 231, 237
top soil stripping 238
topographical plans 180, 181
torsion 71, 75
traffic volume 204
trapezoidal rule 180
truss and purlin construction *169*
two-way supported slabs 45-*46*, 67-69, *68*

U
unequal leg angles *see* angles, unequal leg

V
vertical curve alignments 231-232, 235, 238
volume calculating diagram *213*

W
wall footings 17
web cleats *136*, 149
welds 123-126
    truss detailing *156*, *158*
    intermittent 126, *127*
    sizes of 125
    symbols for *124*, 125-126
    types of *124*
    vs bolts or rivets 155

XYZ
yield stress 98, 100